Science and Fiction

Science and Fiction – A Springer Series

This collection of entertaining and thought-provoking books will appeal equally to science buffs, scientists and science-fiction fans. It was born out of the recognition that scientific discovery and the creation of plausible fictional scenarios are often two sides of the same coin. Each relies on an understanding of the way the world works, coupled with the imaginative ability to invent new or alternative explanations—and even other worlds. Authored by practicing scientists as well as writers of hard science fiction, these books explore and exploit the borderlands between accepted science and its fictional counterpart. Uncovering mutual influences, promoting fruitful interaction, narrating and analyzing fictional scenarios, together they serve as a reaction vessel for inspired new ideas in science, technology, and beyond.

Whether fiction, fact, or forever undecidable: the Springer Series "Science and Fiction" intends to go where no one has gone before!

Its largely non-technical books take several different approaches. Journey with their authors as they

- Indulge in science speculation—describing intriguing, plausible yet unproven ideas;
- Exploit science fiction for educational purposes and as a means of promoting critical thinking;
- Explore the interplay of science and science fiction—throughout the history of the genre and looking ahead;
- Delve into related topics including, but not limited to: science as a creative process, the limits of science, interplay of literature and knowledge;
- Tell fictional short stories built around well-defined scientific ideas, with a supplement summarizing the science underlying the plot.

Readers can look forward to a broad range of topics, as intriguing as they are important. Here just a few by way of illustration:

- Time travel, superluminal travel, wormholes, teleportation
- Extraterrestrial intelligence and alien civilizations
- Artificial intelligence, planetary brains, the universe as a computer, simulated worlds
- Non-anthropocentric viewpoints
- Synthetic biology, genetic engineering, developing nanotechnologies
- Eco/infrastructure/meteorite-impact disaster scenarios
- Future scenarios, transhumanism, posthumanism, intelligence explosion
- Virtual worlds, cyberspace dramas
- Consciousness and mind manipulation

More information about this series at http://www.springer.com/series/11657

Damien Broderick

Consciousness and Science Fiction

Damien Broderick
San Antonio, TX
USA

ISSN 2197-1188 ISSN 2197-1196 (electronic)
Science and Fiction
ISBN 978-3-030-00598-6 ISBN 978-3-030-00599-3 (eBook)
https://doi.org/10.1007/978-3-030-00599-3

Library of Congress Control Number: 2018958318

This Springer imprint is published by the registered company Springer Nature Switzerland AG
The registered company address is: Gewerbestrasse 11, 6330 Cham, Switzerland

For Gregory Benford
Russell Blackford
and Gary Livick
with thanks
And in memory of David G. Hartwell
who got my first novel into print
a shockingly long time ago

Introduction

Until recently, few critics and theorists of science fiction (sf) have examined closely a topic long considered suspect by science itself, especially in the doleful behaviorist years that ended only in the early 1960s or even later. That maligned topic was the nature and structure of the *mind*, its design and variations, and its products in the making of our *selves:* thinking, emotion, *consciousness* itself.

Francis Crick, co-discoverer of DNA's structure and in later decades a brain researcher, cited a cautionary quote from philosopher John Searle: "As recently as a few years ago, if one raised the subject of consciousness in cognitive science discussions, it was generally regarded as a form of bad taste, and graduate students, who are always attuned to the social mores of their discipline, would roll their eyes at the ceiling and assume expressions of mild disgust."

For several decades now, sophisticated emphasis on cognition and consciousness has clambered back toward prominence, but in forms that are still mutually at war. It is the nature of consciousness that this book investigates, although from an unusual angle: how this contested domain of scientific endeavor affected the literary form known as science fiction, and perhaps to some small extent how it was in turn influenced by sf's speculations.

* * *

In the years before most of us were born, science fiction tended to be derided as "that crazy Buck Rogers stuff." Buck is now pretty much forgotten, but in the late 1920s and for some decades beyond, he was a meme crossed with an icon, representing the wonders of a space-faring tomorrow. In 1928, a novella by

Philip Francis Nowlan in the pulp magazine *Amazing Stories* featured Anthony Rogers, a former WWI soldier who is preserved by radioactive dust nearly half a millennium into the future. *Amazing Stories* was the first outlet dedicated to what was then coming to be understood as an entirely new form of popular fiction, first clunkily dubbed "scientifiction" by *Amazing*'s founding publisher Hugo Gernsback and then shaved back to the now familiar "science fiction," sometimes with a hyphen, and abbreviated as "SF" or "sf."[1]

By the end of 1928, Nowlan had adapted his tale into a comic strip, changing Anthony's name to the more swaggering "Buck." There had been plenty of examples of this new genre in print for decades—famously, the slightly futuristic tales of Jules Verne and the brilliantly imaginative stories and novels of H.G. Wells—but Buck caught the wind of a postwar technological surge. Drawn crudely by Dick Calkins, it deployed spacecraft of some ingenuity, a leap of faith well past propellers and jets into a future still not yet here, with home-made rocketships and interplanetary war.

Artist and illustrator Steven Stiles has commented on the impact of Buck's ludicrous, rather adolescent adventures in branding sf of any kind as a sort of children's playground of the mind. "For far too many years we suffering science fiction aficionados were confronted with the descriptive and dismissive phrase, 'Oh, that crazy Buck Rogers stuff!' When applied to the works of, say, Robert Heinlein, Isaac Asimov, or Philip K. Dick, it could be pretty galling. Now the phrase is '*Oh, like Mr. Spock!*,' which is, of course," he added sardonically, "a vast improvement."[2]

Star Trek did indeed drive a simplified form of futurist sf into the minds and hearts and even mini-skirted couture of many TV and eventually movie viewers, from 1966 to the twenty-first century. It spawned an endless sausage factory of paperbacks set in variants of the *Trek* universe—some of this sharecropping surprisingly well written and deftly handled, but all of it crippled by the need to stick pretty much within what had gone before. The same triumph of old ideas handsomely decorated with special effects and scenes drawn from sf magazine covers of the 1940s and 1950s infused the *Star Wars* franchise from 1977 to this day.

Science fiction proper, by contrast, while now often retreated into safe clichéd ruts, has the capacity to generate novelty, not just comic-book inventions like Buck's war cry "Roaring rockets!" and his girlfriend Wilma's flying belt. Sf is able to ask genuinely startling questions and then to pose even more

[1] The term "sci fi" arrived several decades later, a neologism coined in parallel to the then-sexy new "hi-fi" recording medium, hated by the writers of the new medium, but eventually adopted anyway by the public.
[2] http://stevestiles.com/buckrog.htm

startling answers to them. It did so in the mode of empirical and theoretical science rather than, say, fantasy or chrome-plated history reruns. There is a term for this process familiar to philosophers and physicists, perhaps coined by Einstein who was a master of the form: *gedankenexperiments*, or conceptual trials undertaken purely in imagination—mental modeling of situations and outcomes that have never yet happened but can cast a searching light on problems and phenomena that science has not yet resolved.

So the "expert dreamers," as sf writer and editor Frederik Pohl called them, catch the scent of new notions and project possible outcomes that might change the world we know or at least challenge our certainties. Could humans voyage to the Moon, or deep beneath the ocean? For centuries these were enthralling speculations enjoyed by the curious, but by the nineteenth century Verne and others witnessed all around them the innovations of the Industrial Revolution. On this basis they built imaginative fiction (often quite wrong in detail, like Verne's gun-fired lunar ship that would have crushed its crew) embodying this sense of an unleashed future. Decades before President John Kennedy promised to land humans on the Moon, scientifically informed sf writers such as Arthur C. Clarke and Robert Heinlein and illustrators such as Chesley Bonestell were pondering both the visceral thrills of space voyages of exploration and the mathematically precise science of orbital mechanics that made the stories much more than easy handwaving.

As sf developed what I call its *megatext*—the virtual collective encyclopedia of typical tropes, terminology, characteristic plot moves, favorite locations, weapons, tools, psychological shifts, new cultures, etc.[3]—story-making became elaborate yet often non-declarative. That is, grasping the meaning and impulse of an sf story did not require detailed conscious awareness of these narrative devices, because the tropes had become shared commonplaces even as they presented the purportedly unfamiliar—the "shock of the new."

So narrative nudges and an established lexicon lead the reader into somewhat unexpected alternative realities that are at the same time supported by recognition and familiarity. Just as no sophisticated short story writer for the *New Yorker* pauses routinely to explain at length how a car or smartphone or candle works, no sf writer lards the story (as they did in the 1920s and 1930s) with expositional lumps of background. Scene-setting is absorbed indirectly, from the deft placement of new or adopted words and events. So as fashion and core interests shifted from simple space and time travel adventures to florid galactic empires and then more realistic climate catastrophes and pandemics of

[3] See my *Reading by Starlight*, Routledge (1995).

our home planet, and from television and "atom bombs" to the Internet, virtual reality, cyberwar, and life extension, the topics sf dealt with inevitably broached both the predictable new sciences and their plausible findings, both alarming and enthralling.

Much of this history and narrative architecture has been discussed at great and searching length by scholars of the fantastic—of *fantastika*, as the great critic John Clute dubs the generic landscape of this fiction. Russell Blackford has probed aspects of moral philosophy and its manifestations as both fictive background and *gedanken* in another volume in this series, *Science Fiction and the Moral Imagination* (2017). Gary Westfahl drew on the very practical and gritty details of space travel to test how movies and written texts use these icons in *The Spacesuit Film: A History* (2012) and *Islands in the Sky* (1996, 2009) on the space station theme. At an orthogonal extreme, in *The World Beyond the Hill* (1989), Alexei and Cory Panshin looked exhaustively at the "Golden Age" science fiction of the first half of the twentieth century with a developing emphasis that is captured in the subtitle: *Science Fiction and the Quest for Transcendence.*

* * *

Even in these days of advanced instrumented brain and body scanners, with their ability to trace individual neural responses to a given stimulus, the question of what exactly consciousness *is*, and how its special qualities can possibly arise from material flesh and blood and electromagnetic fields, remains what it has been dubbed for some decades: the *Hard Problem.*

Even as the history of the entire local universe has been traced back nearly 14 billion years to the Big Bang, and perhaps beyond if our universe cycles, even as the Standard Model of elementary particles and fields resists replacement as the canonical, immensely powerful account of physical reality, still nobody knows how mind and consciousness fit into the vast panoply of what is known.

You might expect that science fiction has embraced this mystery, this baffling lacuna in the very core of our understanding of life, the universe, and everything, or at least not ignored it. True enough, if usually rather glancingly. Often the thought experiments touching on consciousness are sidelong and, certainly in the early days, rather coarse. In this book, we shall explore both the science and philosophy of consciousness and self and its manifestations in speculative narrative.

One intriguing aspect of the search for explanation is the use philosophers have made of a particular kind of bare-bones thought experiment, typified by the "brain in a vat" trope. It is no accident that these quests after deep truth should be exemplified in playfully simplified yet quite serious mini-sf stories. As philosopher John Searle put it in *Consciousness and Language*:

> The way that human and animal intelligence works is through consciousness. We can easily imagine a science fiction world in which unconscious zombies behave exactly as we do. Indeed, I have actually constructed such a thought experiment, to illustrate certain philosophical points about the separability of consciousness and behavior. But that is irrelevant to the actual causal role of consciousness in the real world. (Searle 2002, p. 29)

French neuroscientist Stanislas Dehaene, in *Consciousness and the Brain* (2014), treats sf's speculations as a shorthand for supposedly outlandish ideas that, surprisingly, are actually now taken seriously:

> . . .late brain activity should embrace a full record of our conscious experience—a complete code of our thought. If we could read this code, we would gain complete access to any person's inner world, subjectivity and illusions included.
>
> Is this prospect science fiction? Not quite. (p. 145)

He says of another uncomfortable notion that "Although this might sound like science fiction, several variants of this experiment have already been performed, with considerable success" (p, 155). Science fiction aficionados need not feel affronted by this common (mis)usage, I think. The way in which people often reach out rather blindly for phrases such as "mere sci-fi" and "sounds like science fiction" is in fact an acknowledgement of how much that is genuinely shocking or apparently unbelievable has already been previewed in sf stories, novels, and even some movies and TV series.

* * *

So what is consciousness, and how can it be captured, or at least pursued, by *fantastika*? As recently as twenty years or so ago, "consciousness" was still a word many scientists dared not utter, for fear of stinging mockery. Now we have become accustomed to reading titles or subtitles from neuroscientists and artificial intelligence specialists: Dehaene's *Other Minds: Deciphering How the Brain Codes our Thoughts*, or Google's Ray Kurzweil's bold (perhaps brash) *How to Create a Mind: The Secret of Human Thought Revealed* (2012). Or Jeff Hawkins' milder title *On Intelligence* (2004) that nonetheless bids to lay out the design for a working artificially intelligent (AI) mind (although perhaps not a conscious one). Or William Calvin's *How Brains Think: Evolving Intelligence, Then and Now* (1997), from closer to the dawn of this new age of consciousness studies. (*Journal of Consciousness Studies* itself, a peer-reviewed and interdisciplinary scholarly publication, only arrived in 1994.)

Of course, consciousness is not restricted to the kinds of skills that are identified in IQ tests. In fact, many of those abilities are more or less *non-*conscious, capable of resolving puzzles fast and intuitively rather than carefully, slowly, and step-by-aware-step. This distinction is elaborated by Antonio Damasio, in *The Feeling of What Happens: Body and Emotion in the Making of Consciousness* (1999), *Descartes' Error: Emotion, Reason and the Human Brain* (revised 2005), and *Self Comes to Mind: Constructing the Conscious Brain* (2010). Damasio—currently the David Dornsife Professor of Neuroscience, Psychology and Philosophy at the University of Southern California, an adjunct professor at the Salk Institute, and head of the Brain and Creativity Institute—discerns the crucial role of non-algorithmic passions and other emotions in learning and acting intelligently. Such emotional aspects of the self contribute to consciousness: both warmth and love that bond us, and the rages that can destroy lives.

* * *

Consider the scenario in the 1942 novel *Donovan's Brain*, by Curt Siodmak, filmed in 1953 under the same title. The brain of a dying wealthy industrialist is separated at the moment of death and stored in a saline solution, alive and conscious. This brain, lacking the organs of conventional communication, discovers paranormal abilities of mind reading and, indeed, the ability to seize control of physically complete persons. A more accurate title of this horror scenario might be *Donovan's Soul*, depending as it does on a dualistic, Cartesian worldview where the mind is coupled to the brain but not identified with its processes.

Tangentially, we might wonder how the kind of "Beam me up!" teleportation common to *Star Trek* and other cost-saving science fiction programs could work. Are the material constituents of a crew member ripped apart and flung through space or maybe hyperspace and then reassembled at the other end? Or is the only "thing" sent and received a package of information, perhaps in binary code, which is then somehow used to reconstruct the traveler from available atoms at the destination? If the Immaterialists are correct, how does the soul hitch a ride on the trip along with the dissociated electrons? This looks like a knock-down argument for emergent physicalism, but I assume the soulists among us would explain gently that the Great Primordial Cosmic Consciousness is already everywhere and everywhen, and its human subunits can plug in anywhere and any time, being built, after all, entirely out of that nonlocal Consciousness. If so, one wonders, why doesn't this happen *all the time* with telepathic messages and clairvoyance? Oh, there's this Aldous Huxleyan valve that shuts out the all-but-infinite downpour of confusing messages from *Everywhere*, except when it doesn't, which is most of the time. Reason, wielding Occam's Razor, has trouble winning in this sort of game.

By the early to mid-1940s, the sf genre saw an explosion of approaches to psychological and social quandaries, some of them quasi-scientific versions of old fairy stories or mythic tales and others drawn from new discoveries and narrative ploys. Ancient legends of possession by demons were transformed into new-minted anxieties about mind control from outside the supposedly unitary self, and personality fractures and multiplicity from within it. Psychiatrists wrote case studies of multiple personality (later known as "dissociative identity disorder"). In the 1950s, non-sf bestsellers such as *The Three Faces of Eve* (1957, by psychiatrists Corbett H. Thigpen and Hervey M. Cleckley) and Richard Condon's 1959 thriller *The Manchurian Candidate* (where the doubling was induced for Cold War political motives) became profitable movies.

Science fiction had already dealt with this topic in notable short stories such as Wyman Guin's "Beyond Bedlam" (1951), where social violence, unhappiness, and war were mitigated by mandatory fracturing of each citizen's mind into two segregated personalities. This novelette shall be discussed in more detail below. Damon Knight imagined the reverse effect, where several people become merged into a single composite ("Four In One," 1953). In a teleportation accident, J.T. McIntosh distributed aspects of one passenger among the personalities of the survivors ("Five Into Four," 1954). Theodore Sturgeon famously and to splendid effect conceived of "gestalt" collectives (*More Than Human*, 1953, and "To Marry Medusa" 1958, expanded as *The Cosmic Rape*, 1958) linked paranormally but retaining an enhanced measure of individual consciousness. We shall look more closely, below, at this problematic, and at science fiction's multiplex explorations of what has come to be known as "hive minds."

A warning, though, before we begin looking at the fiction in detail:

> This book is no respecter of *spoilers*—that is, explicit revelations of plot arcs and character secrets—when such revelations are necessary to make a point about treatments of consciousness in sf, which is most of the time. If you can't bear to know in advance how the plot of a novel, movie or play works out ("Oh my god, you mean *Humpty Dumpty* falls to his *death*!") please read the stories or books discussed below before you engage with any of these chapters. And *enjoy!*

* * *

Consciousness, of course, is so primal, and in human terms so fundamental, that its operation infuses all fiction. In the larger world the word took on specialized color from at least the time of Karl Marx, whose term "class consciousness" registered the sense people had of their status and roles in a differentiated society where work, opportunity, and wealth became shaped by changing technologies. By the 1960s and 1970s, people's growing awareness of such factors emerged in activism motivated by "consciousness raising," a visceral and intellectual realization of the constructedness of the reigning social order, its contingent, potentially mutable, and often oppressive authority. It was especially vigorous in the women's movements, the Black Consciousness Movement in South Africa and then elsewhere, and somewhat later in gay, lesbian, and trans activism strengthened in significant measure by the calamity of the AIDS epidemic.

Strikingly, sf by women in particular enacted this kind of shift in awareness. For example, in 1967, Kate Wilhelm's "Baby, You Were Great" presented a scorching denunciation of the male rape-inflected gaze taken to an extreme using penetration of a victim's qualia—subjective, aware experiences—via a direct virtual reality feed from her consciousness. Ursula K. Le Guin's novel *The Left Hand of Darkness* (1969) brought into imaginative focus gender and sexual fluidity with its planet of unisexual people who alter seasonally into either sex, neither hierarchically privileged. In Joanna Russ's challenging *The Female Man* (1975) variants of a woman in alternative Earths embody and confront restriction or liberty.

These sexual issues had been the basis of earlier novels and stories written as often by men as women, such as Philip Wylie's *The Disappearance* (1951), portraying the consequences of men and women being sundered for a time into separated realities, each moiety required to learn how to cope in a culture lacking the presence of its traditional and biological binary contrast. In Sturgeon's *Venus plus X* (1960), a clade of humans has achieved a kind of utopia of peace and technical advantage through surgically created hermaphrodity. The important black gay writer Samuel R. Delany adapted a kind of science fantasy in deconstructive portrayals of challenging differences (including inter-species sex) in the *Nevèrÿon* sequence and in the far future novel *Stars in My Pocket Like Grains of Sand* (1984). Each of these thought experiments confronted significantly contrasted and often competing orders of consciousness, cumulatively reaching for the state now dubbed "woke," a condition of awakening, illumination, or enlightenment.

* * *

By the 1980s, these trajectories merged with the ever swifter changes brought on by electronic information storage, computation, and communication. Scientists as well as futurists were speaking with renewed seriousness about artificial intelligence (AI), and even anticipating human or posthuman-level machines, already anticipated in Philip K. Dick's dazingly unhinged phantasmagoria. This expectation was dramatized in cynical, noir *cyberpunk* fiction, typified by novels and stories from William Gibson (who projected a parallel realm, *cyberspace*, attained when brains *jacked in* to immensely capable and conscious machine intellects manifesting as literal gods), Bruce Sterling, Rudy Rucker, and a decade later Neal Stephenson, and at first rather fewer women such as Pat Cadigan and Chris Moriaty. Sequences of posthuman future-making fiction were coming from articulate and politically utopian but wide-awake innovators like Iain M. Banks and Ken MacLeod, two Scots of a radical disposition. Banks's novels about the Culture created a slice of the cosmos where human-like (but not human) people did whatever pleased them under the scrutiny and agency of AI Minds. In MacLeod's brilliant *The Cassini Division*, anarcho-socialists confront immensely powerful and deranged posthumans who upload their minds into machines, infest Jupiter, multiply wildly, create a wormhole from the demolished mass of Ganymede, and obliterate the unmodified human use of electronics. Now sf was merging the sociopolitical sense of "consciousness" and the literal hardware level of neuroscience.

Arguably the watershed moment for a neuroscience flavor of sf was the publication in 1990 of the brilliant Greg Egan's short story "Learning to Be Me," which tracks the ontological whiplash of a young man whose organic brain is backed up by an implanted "jewel" (or *dual*). When the jewel falls out of synchrony with the processes of its host brain, the luckless fellow experiences the life of a philosophical zombie, unable to intervene, condemned merely to observe without volition. But is this paralyzed consciousness resident in the brain, or the dual copy? He learns the answer when the brain is scraped out of the skull to forestall the inevitable degeneration of mortality.

Egan's impressive novel *Quarantine* (1992), and several that soon followed, permitted greater density of invention. Even at shorter lengths, however, Egan remained forceful and intriguing, producing cognitive parables rivaling many by Stanislaw Lem. As the authoritative online *Science Fiction Encyclopedia* notes, " 'Reasons to be Cheerful' " (1997) "is a particularly effective story about an adolescent whose surgery for a brain tumor leads to the placing of his own emotional reactions under voluntary control, making him question the roots of volition and affect."[4]

This newly budding branch of the Tree of Story quickly put out shoots of quite challenging and perhaps obscure stories using what came to be called "crammed prose": unapologetically thought out with some rigor and presented in a version of the unabashed jargon of the technical sciences, yet with poetry of its own. Charles Stross drew attention with stories that were combined in his remarkable novel *Accelerando* (2005), which traced the soaring upward curve of what has come to be called the Technological Singularity. (These days he denies that notion has any application to the real world.) Minds were enhanced, uploaded, modified, nano-miniaturized, and fired toward the stars in spacecraft the size of soda cans. It is possible that the book is written by a rather snarky and sardonic cat that undergoes repeated consciousness improvements.

Even more impressive, as we shall see in more detail in Chap. 10, were the novels *Blindsight* (2006) and *Echopraxia* (2014) by marine biologist Peter Watts. Briefly: blindsight is a neurological oddity caused by damage to a critical part of the visual system, preventing the passage of signals to the brain module that registers awareness of the visual field. Despite this genuine loss of the *experience* of sight, information from the eyes does make its mark on the flux of the body's interaction with the external world, allowing the blinded person to walk safely through an obstacle course illuminated by light, though not in complete darkness. Watts' thesis is that the absence of consciousness might be, in certain advanced entities, an evolved benefit. This apparently absurd proposition is demonstrated through utterly straight-faced and terrifying slapstick. It is the kind of *gedanken* that philosophers such as Daniel Dennett routinely deploy, raised to some dizzying mental power.

* * *

By the twenty-first century, many of the award-winning or nominated and impressive sf works in this mode were written by women. Pamela Sargent's "Not Alone" (2006) uses experimental surgery to give a woman an overwhelming, dominating consciousness of what she now "knows," beyond doubt or dispute, is the divine. When finally this procedure burns out the neural circuits conducive to belief, she is stranded for the rest of her despairing life in a desolate Dark Night of the Soul.

[4] http://www.sf-encyclopedia.com/entry/egan_greg

Rachel Swirsky's "Eros, Philia, Agape" (2009) poignantly examines the shifting consciousness of a robot: "Lucian watches. In a diffuse, wordless way, he ponders what it must be like to be cold and fleet, to love the sun and yet fear open spaces. Already, he is learning to care for living things. He cannot yet form the thoughts to wonder what will happen next."

In Connie Willis's antic noir novel *Crosstalk* (2016), Briddey Flannigan, beautiful and stylish, is a minor executive at Commspan, a communications company locked in a race to beat Apple's imminent release of a new smartphone. She undergoes a fashionable minor brain implant linking an aspect of her consciousness with her fiancé's, in a kind of enhanced emotional bond that is meant to convey no messages other than a profound gush of mutual love. What could possibly go wrong?

Perhaps the most intense of these science fictional explorations of non-normative states of consciousness is Elizabeth Moon's *The Speed of Dark* (2002). The mother of an autistic son herself, Moon tells her futuristic tale mostly through the autistic framework of Lou Arrendale who is faced by the threat, or perhaps opportunity, of having his mind "repaired" by a nanotech procedure. Reviewer and novelist Jo Walton captures this sympathetic journey into one of the human-yet-seemingly-alien states of consciousness that Watts' novels would push into the territory of horror:

> [Lou] is utterly logical. . . and he doesn't perceive social signals except sometimes as an entirely learned and intellectual thing. . .. There's a great immediacy to Lou's point of view, and it's all entirely comprehensible, if deeply weird. . .. It's amazing how much of a life he manages to lead, despite how acutely he feels textures and how much he needs a regulating routine. Besides that, Lou sees patterns in the world, patterns that other people don't see, patterns that are really there and help him cope. Sometimes this is just weird, like when he wants to park on a prime number spot, or counts floor tiles, and sometimes it saves his life. . .. It's indicative of how well Moon shows Lou's perceptions from inside that we come to value what he is the way he is and hesitate with him over having his darkness illuminated.[5]

Of course we know much of this already from other novels, and case studies by neuropsychologists such as the late Oliver Sacks, and autistic people such as Temple Grandin. What makes *The Speed of Dark* compelling, like the best sf dealing with consciousness and its Hard Problem of the sources of experience, is its use of a narrative tool of possible innovations that might reshape a

---

[5] https://www.tor.com/2009/02/20/seeing-patterns-elizabeth-moons-the-speed-of-dark/

person's phenomenology, or allow another person to share the experience hidden away inside the skull.[6]

* * *

In 2008, Peter J. Bentley pushed the quest for the roots of consciousness perhaps as far as it can go. His story "Loop" is very brief, hardly more expansive than some of the thought experiments advanced by analytic philosophers. "If only death were the end of consciousness," his narrator laments.

> "Unable to speak, unable to move properly, unable to understand those around me. Not much better than a cabbage, except for the spark of consciousness within. My mind is intact! I'm still here!... My mother is kneeling nearby, looking impossibly young and pretty... Doesn't she realize I am one of the most dexterous neuroscientists who ever lived?"[7]

She does not, of course. He will not acquire those great skills for decades. Then he will die, and awaken again in his mother's youthful womb, and be born, recalling at the age of nearly two Einstein's meditation on life and death and consciousness: "Nobody is ever lost. Each of us will always exist in a given region of space-time, perfectly preserved as though trapped in amber."

He is trapped in this eternal cycle, forever, and with each return to the same womb and childhood his memory shrinks, his consciousness closes its eyes to the reality, he begins again and forgets.

* * *

This brief sample of a few science fictional steps in the quest to understand and manipulate consciousness offers only the barest sense of a hundred years and more of imaginative engagement with the mysteries of the Hard Problem, the brain and intelligence, the self and its physical host or instantiation. Let us now consider in some detail the empirical, experimental, and theoretic history of the mind—the "ghost in the machine"—and thereafter, in considerable greater detail, its parallel in *fantastika*: the fiction of science.

---

[6] Perhaps I will be forgiven for noting that some of my own science fiction has dealt with these issues, from *The Dreaming Dragons* (1980), where the true location of human and other consciousness is located in an ancient archive created by intelligent dinosaurs, to "The Qualia Engine" (2009) in which a hypergenius youth uses quantum computation to enter the raw consciousness of another genetically enhanced teen.

[7] In *You're Not Alone*, ed. Damien Broderick (Surinam Turtle Press, 2015, pp. 67–8).

Contents

1

The Machine in the Ghost

The proper study of mindkind, as the eighteenth century poet Alexander Pope didn't quite say, is mind—and mind's most baffling consequence, consciousness.

That isn't a slur on the human body, on our passions, intuitions, blood-thrilling joys and sorrows. In the twenty-first century it is generally agreed by scientists that mind, when it is conscious, just *is* the body experiencing its rich throng of impressions and thoughts, recalling the past, relishing the present and preparing for the future, and functioning as a contextual node in the local ambiance of other minds. Thus, at any rate, runs the consensus of today's experts in the cognitive and neuro-sciences. Not everyone agrees. Many religions teach that each human is a kind of metaphysical conjoined-twin, brute matter yoked to sublime spirit. Our suffering flesh might disperse upon death, but soul, mind, spirit (not synonymous, but linked in their ineffability) persist and find justice or at least peace.

While there's no scientific evidence for this latter claim,[1] adherents see the strongest proof in experience itself. We know what we are, from the inside. In the language of philosophers, our most humble experience is drenched in *qualia*—the intimate subjective feelings or qualities that no Positron Emission Tomography scan will ever detect: the sweet taste of a ripe peach, the glowing, calming redness of a sunset (not at all the same thing as the energy spectrum of its light). Qualia, it seems, are the very habitation of mind. If so, how can mind

[1] A few scientists and philosophers strongly defend an afterlife (and before-life as well). Different views are canvased in *Immortal Remains: The Evidence for Life After Death* (Rowman & Littlefield, 2003) by emeritus professor Stephen E. Braude. In much of the non-Western world, it is taken for granted.

D. Broderick, *Consciousness and Science Fiction*, Science and Fiction,
https://doi.org/10.1007/978-3-030-00599-3_1

be nothing more than the brainy body in action? On the other hand, cognitive science researchers and evolutionary biologists are gradually showing how mind can emerge unmysteriously from complex neurological structures. Plenty of gaps remain in the chains of explanation, but arguably the story is coming together. Intelligence, awareness and even consciousness have developed—and still function—by Darwinian principles.

* * *

Some, as I say, disagree strongly with this estimate. A professor of philosophy in Santa Cruz, David Chalmers, is an Australian who started in mathematics and computing at Adelaide and Oxford before jumping in at the deep end in Indiana, with computational guru Douglas Hofstadter, to explore mind by traditional and scientific means. For more than two decades he has been at the eye of a storm because of his forthright claim that consciousness is what he calls a "hard problem," that mind has its own special laws, even though it arises from (or "supervenes upon") matter.

In 1997, the celebrated and feisty philosopher John R. Searle, himself a foe of reductive materialism, engaged Chalmers in the *New York Review of Books*, declaring his assertions absurd. Chalmers responded vigorously. Perhaps the most startling aspect of this skirmish is that Chalmers was not a desiccated Jesuit or a bald pipe-smoker in a tweed jacket, but a strikingly good-looking young man with flowing heavy-metal hair. All these years later, with Chalmers' hair much shorter and gray, the debate continues unresolved. Meanwhile, philosopher Daniel Dennett bluntly denies that we are intrinsically different from other kinds of life, even if mind itself is rare. While viruses and bacteria are undeniably mindless robots bustling in our flesh, we must not, Dennett warns, "take comfort in the thought that they are alien invaders, so unlike the more congenial tissues that make up us." Look at them under a powerful microscope and we see beyond argument that the brain's billions of neurons are just cells scarcely different from those of bacteria or the yeast cells in beer vats and bread dough. "Each cell—a tiny agent that can perform a limited number of tasks—is about as mindless as a virus."

Where, then, does mind comes from? From the organization, the networks, the nuanced dance of these specialized ninnies. Not from a mysterious soul, but from the baroque layout of the cellular arrays. Does this mean that dogs also have minds? After all, their brains and nervous systems are not really so much more primitive than ours, compared to bacteria. And certainly there seems to be someone home there, behind the brown loving eyes. Or is that an illusion, projected by the human spirit (which is what we call our wonderful, vastly complex machinery)? But if dogs, what of cockroaches? Ants? Rocks? Dogs, Dennett admits, might be a special case, for they have co-evolved to resemble us in many respects. Yet the case for consciousness in rocks and even elementary particles, seemingly the *reductio ad absurdum* of this line of thought, is increasingly popular among philosophers. We shall return to this bizarre claim a little later.

* * *

William H. Calvin, a Seattle neurophysiologist and polymath writer on science, argues that consciousness—or intelligent awareness, at least—arises because brains are "Darwin machines." Stimuli from the outside world (or from memory) trigger specialized parts of our brains into cloning temporary representations of things we've seen or thought before. The cloned activation patterns multiply in hexagonal arrays of cells half a millimeter apart, hungry for brain-space, synchronizing, reinforcing or inhibiting one another. Each is a kind of fragment of a thought or recognition. Calvin's suggestion is astonishing, testable, and links minds to the rest of the evolutionary universe. Meanwhile, Chalmers remains unconvinced. One can imagine a world of zombies, he asserted, just like ours but lacking consciousness. The brains of these zombies mimic ours precisely, but there's no light on inside. They act and speak and laugh and "love," but are mere automatons. Since this nightmare is logically possible, Chalmers says, consciousness must be something over and above mere neural structure in action.

What is it, then? Information, he concluded. Not an answer to give the religious faithful any comfort, admittedly, but developed since 2004 by neuroscientist Giulio Tononi into the most prominent theory of mind, based on the proposition that the degree of consciousness can be matched against a calculus of available "integrated information" (II, or *phi*). Chalmers, like Tononi later, argued that a sufficiently complex artificial intelligence system would be conscious. So he was not proposing to reinstate an immaterial soul. He put it neatly: "Experience is information from the inside; physics is information from the outside."

Inside what, though? Inside the mind, with the qualia. But that leaves us where we came in. Besides, I would argue that we cannot truly imagine a zombie world, any more than we can truly imagine a world exactly like ours, full of jittering molecules but without heat. Ian Stewart and Jack Cohen provided an amusing analogy to support this suspicion. Imagine a zombike, they suggested, "which is exactly like a bicycle in every way except that it does not move when the pedals are pushed. Oh, mystic miracle of ineffable immateriality, the source of motion in a bicycle is not anything physical!" The zombie analogy, at root, is no more persuasive—nor even, I'd claim, intelligible.

* * *

During the last half-century, cognitive science has rewritten our understanding of the mind. Decades ago, Tom Wolfe declared in Forbes business magazine, of all places, "Neuroscience... is on the threshold of a unified theory that will have an impact as powerful as that of Darwinism a hundred years ago." We live in an age, Wolfe noted ruefully, in which it is both impossible and pointless to avert our eyes from the truth revealed by the subtle new brain scanning technologies. Even so, somewhat to its embarrassment, brain science and computational disciplines are still unable to create an artificial intelligence. Arguably, we need to test how consciousness works (human and otherwise); as noted above, practical theorists such as Google's chief engineer Ray Kurzweil and Jeff Hawkins have been pursuing just such research. Yes, fabulously fast computers can now beat the best human players of chess or Go, but they are seemingly as far from Artificial *General* Intelligence as ever.

So let us glance, modestly, inside the human mind. Or do I, after all, mean "brain"? Here's a starting point: minds are the way the brain minds the body (including the brain itself).

Despite the traditional aura of mystery and reverence, the human mind can seem a vulgar instrument, or at least mine does all too often. It's a silly thing, avid for the slightest punning chance to torment one word into another so I can fall about laughing. And the indecency is only worsened by the pitch of abstraction at which it occasionally vibrates.

The power of verbal categories over our illusory sense of direct, unimpeachable perception of the world is known in linguistics as the Sapir-Whorf hypothesis, now rather in eclipse. Learning of this wonderful conjecture from science fiction magazines was one of the high points of my adolescence. My God! I thought, dazed. Does this mean that the experienced world is radically mutable, and by so simple a thing as language? Our inner world was shaped and constrained by the *words* available to our use! Some years later I mentioned this theory to an urbane academic friend. "Whorf's hypothesis?" he muttered, hardly impressed. "'Ontology recapitulates philology'." I burst out laughing. Let me explain.

I couldn't believe I'd heard him correctly, which happens a lot when you entertain Benjamin Lee Whorf's hypothesis. I supposed that he'd said (though I couldn't imagine why), "Ontogeny recapitulates phylogeny." That is the spurious nineteenth century Biogenetic Law promoted by Ernst Haeckel to account for the way embryos seem to pass successively through the stages of their evolutionary history. And then, in a mental flash of what Arthur Koestler called bisociation—slamming together two disparate frames of reference—I was hooting in purest delight. *Ontology*, the science of what exists, reprises or echoes *philology*, the science of utterance.

I had been surprised by my own mind.

How is that possible? Am I not my mind? Is my mind not me (not I)?

Of course I am my mind, and the body it's running (that is running it), but I am not privy to all its contents and operations. What startled me was an insight forged inside a partitioned-off and inaccessible (or "unconscious") part of it, a pun-making and pun-recognizing gadget that sits there poised like a mousetrap in the folds of the linguistic cortex on the left side of my cerebrum, waiting for a chance to snap shut on some wriggling word.

After all, how do we think up whimsical gags? We prime ourselves, get in the right mood (whatever that is), and out comes. . . something funny. Mildly funny or hilarious, witty or worth a groan. Either way, new! Amazing, really, in a universe of biological machines.

We each have a complex set of schemata—interpretation machines—buzzing away in various not-quite-conscious regions of our brains/minds, competing for attention. They generate structures, and oppositions to those structures, and throw in forgotten quotes, all that routine mental business. I think I think, therefore I think I am. A science fiction writer's best ideas and phrases come out of inaccessible regions of the mental machinery, perhaps even more markedly than those of other people.

* * *

And that's a prime clue to the nature of brain and mind alike. Years ago, Daniel Dennett encapsulated this organizing principle in a prosaic desktop computer phrase: he favors a "multiple drafts theory of consciousness." Mind is not an angel tragically trapped in flesh, but a text-processing program. What You Think Is What You Get. HTML hypertext for the World Wide Mind. That kind of thing. Of think.

To tell the truth, I'm in two minds about Dennett's claims to have solved the mystery. Or do I mean two minds are in me? Maybe more than just two minds are scurrying about inside my head, since I seem to be quite unsettled in my opinions. Yet "opinions" aren't the same as "minds," are they? Or am I confusing "minds" with "selves"? But "selves" have "bodies" as well as "minds," don't they? So should that be "are" bodies and minds, rather than "have"? Or is that true only until the body dies?

Dennett's playful philosophical method, which I'm caricaturing in the previous paragraph, makes a delightful contrast to the preening obscurities of its most notable rival doctrines, Heideggerian phenomenology and its Anglo-French heirs. Dennett and his anti-immaterialist colleagues have been beavering away at the problems of the contents of consciousness, will and intentionality since the mid-1960s, updating their formal arguments in line with floods of research issuing not just from the philosophy seminar but more importantly from the neuroscientist's scanning instruments and the cognitive psychologist's test room. With Douglas Hofstadter, Dennett co-edited an early accessible anthology on the topic, *The Mind's I*, drawing upon science fiction and artificial intelligence research as well as Anglo-American epistemology (the study of how we know things). His approach to the secrets of the self is fairly summarized in the charming but quite serious gag by Hofstadter, author of the formidably interdisciplinary *Gödel, Escher, Bach*: "Is the soul more than the hum of its parts?"

Like a computer humming happily to itself, accepting information from the outside world and processing it within a suite of programs that in turn derives from that outside world, does the brain "run the software" of the mind? Or is this functionalist formula just the most recent in a line of opportunistic analogies, a desperate but doomed attempt to shoe-horn the ineffable into whatever current box of tricks (clockwork, steam engine, electronic gadget, chaos theory, computationalism) is borrowed by eager and shameless reductionists? That's an objection that rose again into philosophical view due to the efforts of enthusiasts like Chalmers for the "hard problem." What exactly can consciousness be? What are experienced qualities, in a world where, science assures us, nothing but energy fields and perhaps information truly exist? It's a crucial question that will be met head-on (so to speak) by the mid-twenty-first century, as we begin to deal with issues like artificial general intelligence.

* * *

Recall that René Descartes, more than three and a half centuries ago, decided that the world was a machine, with one exception: the human mind or soul. His meditations led him to conclude that this non-material entity must be anchored somehow to the rude physical flesh, presumably somewhere in the head. The pineal gland, neatly in the middle of the brain, seemed a good candidate for mind-body docking. Nobody takes that pineal fancy seriously today, but an underlying Cartesian dualism of this sort remains widely persistent. It is supported by an illusion we almost inevitably fall into, that there's an "inner person" behind the eyes, somehow immune to age, continuous despite bodily changes, morally responsible even when moods sway us (for moods seem more "physical" than trains of thoughts, don't you feel?).

Dennett has dubbed this illusion the "Cartesian Theater." With a barrage of superbly chosen clinical and logical examples, he has repeatedly demolished its claims on our intuition. Oliver Sacks (whose most celebrated patient mistook his wife for a hat) helped prepare the ground for us to accept this sort of argument. Lay readers enjoyed Sacks's beguiling case histories; he was, you might say, the neurologist from the Lake Wobegon Hospital. Sacks showed repeatedly that almost any given aspect of what we consider integral to selfhood can be damaged with great precision when brain or bodily function is hurt. The soul's hum falters.

An astounding report of this kind is from a 1983 Munich clinical paper on a woman who is motion-blind. She finds pouring tea tricky, because "the fluid appeared to be frozen, like a glacier." Crossing the street is scary: "When I'm looking at the car first, it seems far away. When I want to cross the road, suddenly the car is very near." Her brain edits out the intervening motion, turning ordinary experience into a lethal video clip. And yet the dignity of the human person need not be compromised, let alone denied, in this kind of "debunking" demonstration. Sacks's wounded patients break our hearts, but we are left with insight rather than a gush of sentimentality.

If we remove the homunculus (or shrunken inner observer) from the imaginary cerebral theater, what are we left with? Dennett's word-processing model proposes mind as Multiples Drafts of a document generated, edited and acted upon in a massively distributed parallel computer. This seems at once vulgar (mind as business secretary) and ineffective (for surely an inner consciousness is still needed to "read" the "document"?). Dennett's triumph has been to lead sceptics by small stages, each modifying yet adding to its predecessors—like a series of drafts of the argument, you might say—until we confront counter-intuitive conclusions that seem altogether convincing.

Is such facile ease a reason in itself to distrust a theory that, like a magical trick, defies our expectations? "I don't view it as ominous," Dennett declared, "that my theory seems at first to be strongly at odds with common wisdom... The mysteries of the mind have been around for so long, and we have made so little progress on them, that the likelihood is high that some things we all agree to be obvious are just not so."

* * *

It is difficult even to hint at the huge amount of evidence and argument cognitive scientists and philosophers have marshalled lately in support of this kind of account, patiently reported by the French neuroscientist Stanislas Dehaene in the fine book *Consciousness and the Brain* (2014). Ranging from very peculiar sensory illusions routinely evoked in the psych lab, all the way to pathological "multiple personalities" found in the victims of horrendous child abuse, they cut the ground from underneath any intuition we retain that mind and self are clearly unitary and transparent to inspection.

Perhaps the most striking example is an experiment in which a red and a green spot of light flash in rapid alternation against a dark background. How would you expect to see this sequence? It's the principle on which television and computer monitors and smart phone displays are based, after all. When we sit in our living rooms or study desks, watching the small screen, colored dots flicker, weaving a sequence of static pictures that our minds blend into the illusion of people in motion. Still, this red/flick/green lab experiment is very weird. The red dot looks as if it's moving across the screen, and then abruptly switches to green in the middle.

Think about it. The thing changes color before your consciousness is aware of the green spot! The Cartesian Theater model tells us that the inner watcher must have been prevented from seeing this sequence until the whole thing was over, whereupon a kind of retrospective censorship or rewriting of history got foisted on us. Dennett has argued lucidly that this is the wrong story to tell. In fact, the observer's point of view is dispersed, "spatiotemporally smeared all over the brain." There's no single inner watcher to be fooled. We are composites, spread out slightly in time and space. Perception is not merely prone to illusion; it's based on it.

Mind, then, is a virtual machine, built by culture, running on the brain's neurological processing system. This is a holistic theory, of sorts, but with quite specific mechanisms and predictions. Perhaps Dennett's strangest suggestion, to those interested in AI and the science fictional prospect of uploading a living mind into a machine implementation, is that the mind's "virtual machinery" uses a linear architecture like today's supercomputers, though it is obliged to run on the brain's quite unsuitable parallel machinery. The alternative reductionist explanation advanced by Hawkins and Kurzweil claims the reverse—that connectionist or neural net data-processors are the wave of the future, because they more closely resemble the human cortex.

Beyond such specialist arguments, though, is the question that keeps me in two minds about Dennett's much-mocked claim to have solved consciousness by explaining it away. Has he, after all? Don't these Multiples Drafts still feel from the inside like a unified self? Don't other people present themselves with a certain fundamental and compelling unity? Of course, admits Dennett. It's because humans have evolved as story-tellers, just as beavers have evolved as dam-builders. A side-effect of grammar is "to (try to) posit a unified agent" speaking these words, "to posit a center of narrative gravity." Just as a body's mass is dispersed throughout its volume, but its motion is governed by an imaginary point, its "center of gravity," so too with us as experiencing persons. "We build up a defining story about ourselves, organized around a sort of basic blip of self-representation," like the blip on a radar screen. "The blip isn't a self, of course; it's a representation of a self."

One amusing aspect of this neuroscience story is that it closely mimics the account evolved by poststructuralist philosophy, semiotics and psychoanalysis. Those uptown intellectual boulevardiers—with their difficult and frequently derided jargon of antihumanism, subject positions, discourse formations, deconstruction and dissemination—turn out to have an eerie resemblance to Dennett's down-home empiricists. Dennett once acknowledged this in a nicely comic moment, citing (with "mixed emotions") a version of his theory he found in David Lodge's campus parody *Nice Work*, attributed to a fashionable English Department deconstructor. Significantly, Dennett's project was hailed by the late pragmatist Richard Rorty, a notable philosopher at home in both these divorced traditions. So while there is no doubt that Dennett's version of self is just one draft in an evolving discussion, it has every prospect of continuing to play a role at the debate's center of gravity.

* * *

Cognitive science is often regarded by humanities specialists as a playing field for deluded positivists whose search for computerized artificial intelligence, necessarily doomed in advance, proves how derrière-garde their gung-ho enterprise must be. A skeptical estimate of cognitive science is not altogether unjustified.

The balance between a kind of holism ordained by any massively-distributed theory of the brain, and the reductionism required to investigate its properties at close range, bedevils any discussion of these issues. We shall find ourselves returning to this background dispute, this nagging tooth of methodology. Consider the case of the late Lewis Thomas, a gifted essayist and former director of the famous Sloan-Kettering Cancer Center in New York. Thomas wrote eloquently for 20 years in the prestigious *New England Journal of Medicine* about whatever took his fancy, which was rather a lot: medicine, of course, both bedside and high-tech; immunology, cancer research and cell biology especially; ecology; the roots of words. His first collection, *The Lives of a Cell*, bowled me over in the 1970s. Musical and learned, gentle-voiced and tough-minded (rather like Oliver Sacks, in some ways his British counterpart),

Thomas craved the global generosity of holistic approaches while admitting candidly that uncompromising reductionism remains the successful method of choice in science.

A recurrent theme in Lewis Thomas's essays was biologist Lynn Margulis' wonderful argument that the complex cells from which we are made are themselves colonies. The mitochondria that power life have their own DNA, passed down only through the maternal line, and seem to be relics of a symbiotic partnership with larger protocells when the world was much younger. So too, Thomas mused, might we stand within the economy of Gaia, our planet viewed as a self-sustaining quasi-organism. Here he ventured close to gibberish, but it might prove to be the kind of gibberish we need to get us through the stresses of the twenty-first century (and it formed a key pillar in Isaac Asimov's concluding volumes of his *Foundation* sequence). Thomas wrote:

> Given brains all over the place, all engaged in thought, and given the living mass of the earth and its atmosphere, there must be something like a mind at work, adrift somewhere around or over or within the mass... It is, if it exists, the result of the earth's life, not at all the cause. What does it do? It contemplates, that's what it does.

No hypothesis could be more repugnant to Nicholas Humphrey, a lucid research psychologist and colleague of Dennett. Strictly materialist, he once accepted a fellowship funded to study psychical phenomena—so as to learn why otherwise rational people hold to such doctrines as psychic phenomena and post-mortem survival. It did not apparently occur to him that some of these phenomena, however odious they seem to many scientists, are well supported by laboratory as well as anecdotal evidence.[2] The same easy dismissive tone is evident from Professor Dehaene: "In order to verify [patient reports of out of body experience near death], some pseudoscientists hide drawings of objects atop closets, where only a flying patient could see them. This approach is ridiculous, of course" (Dehaene 2014, p. 45). Knowing that this is true in advance, he feels no need to investigate cases where a patient *has* reported information not available to the senses.

A disembodied mind is, for Humphrey, a contradiction in terms, since the mind is exactly the activity of a brain in a body built by evolution to deal with the teeming reality of the physical world. Like Dennett, Chalmers and Dehaene, he pursues the toughest of all assignments, a satisfying account of

[2] See Broderick and Goertzel, 2015.

consciousness. For even if the structure of neurons and chemical transmitters is clarified, even if PET and MEG scans literally show a thought passing through a brain, how can a sack of atoms add up to *feelings*, to sensations and awareness?

His answer is simple enough: consciousness is sensation—not "perceptions, images, thoughts, beliefs"—become self-aware in feedback loops inside the brain. (Hence, unlike Chalmers and others, he insists that robots will never attain consciousness, since they will lack the evolutionary history of our senses.) Sights feel different from sounds and touches due to an evolutionary shaping process, because we need to tell these channels apart in an instant, for our survival. Feelings are "activities that we ourselves engender and participate in—activities that loop back on themselves to create the thick moment of the subjective present." Like all the arguments we are glancing at, Humphrey's rich case can hardly be summarized so curtly. Still, while I sense that it's persuasive enough, I perceive, think and believe that, in exalting sensation over the rest, he may be too hastily in flight from the reigning computational accounts of mind.

* * *

Perhaps we too easily assume that people are utterly different from machines, even from today's Go-mastering neural networks. Well, don't we do something no machine ever displays: break free of set patterns, act creatively? Margaret Boden, research professor of cognitive science at the University of Sussex, with degrees in medical sciences, philosophy, and psychology, denies this. As deft with musicology, romantic poetry and the history of science as she is with computational psychology, Boden gives such complacent prejudice short shrift. Creativity, she observes, implies a capacity to jump free from a set of constraints into a surprising "impossible" solution. To a modest extent, computer programs had already done this decades ago. They rediscovered laws of physics using the rude data and heuristics available to scientists in centuries past. Computer programs have written passable jazz (though not convincing poetry), and found unexpected solutions to mathematical puzzles.

I once wondered idly if our fascination for black holes in space might include a kind of accidental hard-wired neurological explanation. A recurrent nightmare of mine as a small child was both terrifying and non-representational; I have always assumed it tapped into some primary perceptual building-block schema, the kind of cognitive template we construct our subjectivity from. In my nightmare, I was in the presence of unavoidable shapes like infinitely elongated cones, black and awful, curving downward sharply from a flattened top. They looked, I much later realized, like embedding diagrams of black holes in hyperspace...

* * *

Neuroscience is typically done "bottom-up," following the assumption that big things are built from little things, and take much of their form and function from the constraints imposed at the lower levels of organization. On the other hand, there has been a huge shift in emphasis during the last two decades. Plenty of folks working in the neuroscience labs concluded that *complexity* and *emergence* were the watchwords for their developing understanding of self and its gooey components under the skull. Although I am impressed by the computational model, the best explanations will transcend metaphors based on rigid styles of programming in which information is sent through a processing unit in a stream of well-defined tasks. John McCrone observed, by contrast, that

> modern neuroscience sees the brain as a dynamic neural network. A "state of information" has to grow organically, evolving under the pressures of positive and negative feedback until it reaches a state of balanced tension. In such a network, it is not the speed of traffic along the individual wires that counts but the performance of the entire network as it settles into a "solution state."

This does not mean we must revert to the Cartesian Theater, that mythical site deep within the brain (or outside it, in the "soul'). Rather, it implies that thought and awareness resemble the self-organizing—or, better, mutually-organizing—patterns created by the solar system as its planets and asteroids and comets orbit our central star, the Sun, falling into typical "attractors" or balanced orbits while remaining open to perturbation, especially from the changing geometry they create with respect to each other.

No immaterial machine calculates these dynamical patterns, they simply emerge from the physics of space and time. Yet we can mimic those patterns by cunning and insightful equations that evolve with lightning speed inside a computer. Perhaps the way the brain works is like that: it just settles into place after a disturbance, as sand piles up into a cone when it is trickled from above. No computation is needed. And out of these astonishingly ornate and hidden orbits about the attractors of the mind, a fairly continuous "self" is contrived.

But is it sensible to speak of AI machines as "artificial brains"? However powerful their modular structure grows, however advanced beyond the robotic kitten or even talking head stage (already a quite staggering achievement), will they really have the capacity to host an artificial mind, or even an artificial self?

* * *

I attended an academic conference on mythopoeic literature titled "Desperately Seeking Selfhood." While the room had its share of pre-deconstructive academics and Jungians whose notions of wild transgression seemed confined to fairytales and *Star Trek*, many papers tore at certainties of identity, of self. Nor was this accidental. Far from seeking selfhood, the main projects in both humanities and life-sciences today desperately flee from it.

Or so it might seem. In fact, that was chiefly true of the poststructural insistence on the decentered and disseminated self, the doctrine of self as unauthored text initiated by French theorists Louis Althusser, Jacques Lacan, Roland Barthes, Jacques Derrida and Julia Kristeva. Self, they argued, was largely "written" into existence by impersonal culture, rewritten by every reader—every other human—encountering it. (But who are these readers—for their selves, too, must be just as volatile and uncertain?)

Cognitive science held the patents on most of the mind paradigms in play outside the humanities departments, and it and neuroscience in turn, as we have seen, slowly picked apart the self into a vast flow-chart of dedicated modules, of specialized mental organs and their distinctive tunes. Oddly, unlike poststructural accounts of the self and its representations of body and world (the kind Daniel Dennett found so reminiscent of his own work), this story found a certain fondness for individuality, systematic emergence, an information-integrated "self" implicit in each unique genetic and epigenetic program.

Consider this horrid but informative experience suffered by Chris Langton, the brilliant oddball who almost single-handedly invented what came to be called A-life or "artificial life": computerized simulations of living processes. Hang-gliding some decades ago, he botched his landing and smashed into the ground, breaking most of his bones and badly damaging his head. His description of what it felt like to come back on-line as a person is extremely provocative:

> I had this weird experience of watching my mind come back... I could see myself as this passive observer back there somewhere. And there were all these things happening in my head that were disconnected from my consciousness. It was very reminiscent of virtual machines... I could see these disconnected patterns self-organize, come together, and merge with me in some way... It

> was as if you took an ant colony and tore it up, and then watched the ants come back together, reorganize, and rebuild the colony.[3]

These are suggestions rather more rigorous and at least as interesting and fecund as, say, those of the poststructuralist psychoanalyst Jacques Lacan (still somewhat fashionable although he died in 1981, and his key work had been finished long before that). Langton's detailed account of his mind re-booting is profoundly suggestive to anyone toying with the notion that selfhood is comprised of multiple subsidiary "minds," intelligences, modules, and agents, mostly inaccessible to consciousness. But we can't help asking: who was the me with whom these tiptoeing mental fragments of the old Langton were merging? Was it just the sketch of the "full me" the rest of us have, or, more accurately, are? Or was his "self," and ours, a sort of memory-limited buffer, or working space, where subordinate modules might dump their partial comments about the world, to be written together there into the text of the experienced self?

"So my mind was rebuilding itself in this remarkable way," Langton recalled later.

> And yet, still, there were a number of points along the way when I could tell I wasn't what I used to be, mentally. There were things missing—though I couldn't say what was missing. It was like a computer booting up. I could feel different levels of my operating system building up, each one with more capability than the last. I'd wake up one morning and, like an electric shock almost, I'd sort of shake my head and suddenly I'd be on some higher plateau. I'd think, "Boy, I'm back!" Then I'd realize I wasn't really quite back. And then at some random point in the future, I'd go through another of those, and—am I back yet or not?. . . When you're at one level, you don't know what's at a higher level.

Now of course, this report can't be taken, at face value, as proof that the mind is discontinuous or modular, let alone that it's organized like a linear computer program, using some kind of familiar systems architecture built out of suites of sub-programs, able to call specialized sub-routines. In fact, we are pretty certain that the human mind isn't very much like that at all. If it has a computational basis, its architecture will be massively parallel rather than linear, despite Dennett's claim that we function in a linear if weirdly distributed fashion. What's more, Chris Langton was on industrial doses of drugs, so their impact, coupled with his own interest in computation, might have

[3] Cited in Waldrop, *Complexity*.

produced a confabulated construct in his damaged brain, no more veridical than the strange experiences reported by people who are certain that they've been abducted by UFO occupants.

Nevertheless, Langton's testimony fits in very evocatively with other material that has surfaced from neurological case studies and neuroscience investigations of how both normal and abnormal brains appear to function.

* * *

Happily, there's been a growing shift in emphasis. This tempers the stern physicalism of Francis Crick's so-called "Astonishing Hypothesis": namely, "that each of us is the behavior of a vast, interacting set of neurons," and—culture aside—nothing more. Crick deliberately restricted his own post-DNA research to the problem of vision, which is more tractable than seeking wildly for a general theory of mind. What is the path, though, from the retina to our awareness of seeing? Ray Jackendoff, a linguist influenced by MIT's Noam Chomsky, suggested that consciousness dwells neither at some integrated level of the entire cortex nor at the lowest level of stupid individual neurons. The MIT philosopher Jerry Fodor, whose important texts include *The Modularity of Mind* and *Psychosemantics,* insisted that our (surmised) central interpreters are unlike modules—being dispersed and non-localized, in some way mysterious and holistic. Jackendoff was convinced that even the inner interpreters are partitioned.

How is it, then, that we co-ordinate our representations of world and self? We do so through abstract mental models. These are never available to conscious awareness, but we can be sure they must exist as the templates for our neural computations. We never see things in true three-dimensional form, for example, since that would require us to observe, in the same instant, the back and front and top and bottom and insides and spatial orientation of an object. But we certainly construct interior models of the world with this rich 3D character, and it's in these abstract models that different sensory modalities—sight and touch, say—are brought into common registration.

* * *

Strange tales in support of this view popped up in neurological studies of both normal and abnormal brains. Fans of nineteenth century poetic symbolism and impressionist painting will recognize "synesthesia," a curious condition where one sensory channel seems to be cross-wired to another. Sixty years ago, novelist Alfred Bester's vivid protagonist Gully Foyle was slammed into synesthetic confusion in *Tiger! Tiger!*, a classic sf novel serialized in *Galaxy* magazine as *The Stars My Destination*:

> Touch was taste to him... the feel of wood was acrid and chalky in his mouth, metal was salt, stone tasted sour-sweet to the touch of his fingers, and the feel of glass cloyed his palate like over-rich pastry... Molten metal smelled like blows hammering his heart...

Blind to its declarative crudity, I was beside myself with rapture when I read this at 14 or 15, especially since my own mental-imagery repertoire is almost non-existent. I can't even make a picture of a red triangle in my head, let alone one that tastes like a lobster. One genuine synesthetic, Michael Watson, was repeatedly brain-blood-flow scanned and drugged and tested. Watson specialized in linking odor/tastes with rudimentary but intense tactile impressions. The knack was beyond his control. Removing a chicken from the oven, Michael would be distressed if it had too few "points"—it tasted too "round." Sucking spearmint under lab conditions, with amyl nitrite to enhance his sensory cross-link, and then amphetamines to dampen it, he literally felt the presence of smooth, glassy columns, projected outside his body.

Only one person in ten million is a full-blown synesthete, according to neurologist Rick Cytowic, who studied Watson. Cytowic started by investigating synesthesia and ended with a new model of the brain. High-tech tests suggest that we all make these translations from one sense impression to another, but it occurs at a computational neural "level" prior to consciousness, and usually hidden from it. Cytowic tracked the activity to the limbic system, deep inside the brain. It only comes to consciousness when high cortical activity is, so to speak, switched off. Usually this happens only to people with disagreeable brain damage and consequent terrible deficits of awareness or ability. Rare, healthy synesthetes like Michael Watson enable us to peek in and see the way our normally-hidden or protected brain processes are partitioned and/or overlap.

Few topics are more enthralling, and infuriatingly evasive, than the brain and its workings. The human mind is supported by a brain that is parallel and multiplex in function, distributed rather than localized, with a reality-mapping cortex but a limbic zone which "determines the salience of that information." So a powerful emotionality lies at the heart of our humanness. It might well be that synesthetes gain access, otherwise-forbidden (except in "out-of-body" experiences), to the processing of the 3D "model world." This might be conducted in the hippocampus and other regions of the emotional and memory centers of the limbic system devoted to evaluating (or labeling) the salience, the relevance, of what we experience. Synesthetes gain that forbidden access precisely by reducing the blood supply to the cortex, and flooding the deep core of the left hemisphere with rich energy supplies. Using modern scanning devices, you can see it happen in real time on the screen.

For Cytowic, this is proof that the self is primarily emotional rather than rational, located in the deep brain rather than the neocortex that covers it like crushed gift wrapping. Even more piquantly precise was Crick's suggestion that Free Will is localized in a portion of the inner brain, near the top and toward the front, called the anterior cingulate sulcus. Damage to this small group of cells is known to have caused an otherwise alert patient's mind to become "empty," unable to communicate but unworried by that awful loss. It is a notion to cheer the shade of Descartes with his belief that the soul was attached to the brain at the pineal gland—and perhaps just as daft.

* * *

The same general case was advanced by Michael Gazzaniga, the neuroscientist who taught us decades ago about "split-brain" patients. Gazzaniga argues forcefully for a modular account of both specialized brain circuits and, more explicitly, of the sources of consciousness. He drew his theoretical inspiration from Jerry Fodor, philosopher of modularity. In the first place, he showed that right and left cerebral hemispheres possess (on the whole) quite different talents. We seemed to blend at least two minds, joined by the passage of information via the corpus callosum: one verbal and logical/interpretative, the other spatial and intuitive. This account, we notice at once, nicely confounds the ingredients often supposed to segregate male versus female. On a more detailed account, our brain's specialized modules (rather like the "faculties" of an earlier philosophy) interact through an "interpreter" region.

It looked very much, as Gazzaniga states, "that the brain is indeed organized in a modular fashion with multiple subsystems active at all levels of the nervous system and each processing data outside the realm of conscious awareness... These modular systems are fully capable of producing behavior, mood changes, and cognitive activity." Other experts thought this account was not the whole story. William Calvin and linguist Derek Bickerton declared in 1999: "We have shown language to be innate, species-specific, supported by task-dedicated circuits, even if parts of those circuits may do double or treble duty in other tasks"—while adding that "brain imaging has shown that strictly locationist models of language function won't fly... But then it was Fodorian psychology rather than linguistics that insisted on strictly localized, encapsulated modules."

Gazzaniga's work asserted that our specialized brain architecture, which controls what we can know and do, is largely pre-set by the genomic instruction set that ordains each brain's initial wiring. He offered a Darwinian analogy of selection rather than instruction for the process of human learning. Nobelist Sir Macfarlane Burnet's famous explanation of antibody formation showed how a thousand natural shocks prune or select a vast pre-existing array of immune cells. Just so, says Nobel winning immunologist Gerald Edelman, each person's experience sculpts the wild forest of prenatal brain cells into the adult's neat garden of mind. We are born neither "blank slates" nor pre-programmed. Mind resembles the immune system rather more than it does a computer waiting to be loaded with data. From infancy we contain billions of variant antibody molecules, each primed for its antigen of choice. We don't respond to the world's impacts by devising cunning new weapons of defense; we are limited to the armory that we have carried from childhood. Adapting the principles of species evolution to the individual developing brain, Edelman's "neural Darwinism" or "topobiology" tried to explain consciousness in one mighty leap.

His audacity was breathtaking, for he had a strong opinion on every stage of the transition from brute matter to consciousness. From quantum theory to networks of nerves, from "re-entrant maps" that chunk groups of nerves, all the way up to language acquisition—oddly, he does not think highly of Chomsky's suggestion that we have a in-built "grammar organ"—Edelman constructed a testable, integrated model. Gazzaniga admits that while "the functioning modules do have some physical instantiation"—that is, you can run a PET scan or a magnetic resonance, or cerebral blood flow techniques, and actually pin-point the parts of the brain active during a given mental act—"the brain sciences are not yet able to specify the nature of the actual neural networks involved for most of them." William Calvin's intriguing approach attempted to answer that requirement, for arguably brains are Darwin machines not just at the level of neural growth but in their operation as well. In Calvin's account, concepts fight for survival as quite literal feuding activation patterns dispersed across the brain's neural nets.

It seems a fair estimate that none of these models has been generally accepted by the relevant sciences, somewhat as the variety of interpretations of quantum theory have not yet yielded a knock-down winner. It might be that we shall see real artificial intelligence—even general intelligence marked by true experience and awareness—before the scientific community settles on an explanation for consciousness.

2

The Language of the Mind

How can we have a unified and theoretically satisfactory account of ourselves and of our relations to other people and to the natural world? How can we reconcile our commonsense conception of ourselves as conscious, free, mindful, speech-act performing, rational agents in a world that we believe consists entirely of brute, unconscious, mindless, meaningless, mute physical particles in fields of force? How, in short, can we make our conception of ourselves consistent and coherent with the account of the world that we have acquired from the natural sciences, especially physics, chemistry, and biology?. . . I think this problem—or set of problems—is the most important problem in philosophy, and indeed there is a sense in which, in our particular epoch, it is the only major problem in philosophy.

John R. Searle, *Consciousness and Language, p. 1*

Perhaps the most important result of these many and varied empirical and theoretical findings is that human consciousness, together with its predecessors and components, is in certain important ways localized as well as global, just as the cells and functions of the rest of the body chunk together into relatively autonomous organs. Even when the world comes at us in heavily pre-processed human language, we do not always find it easy to comprehend.

It is much worse when we try to understand some completely new aspect of the world, one not yet modeled by our customary fallible set of linguistic gadgets. Arguably, this is why science took so long to emerge, why it has done so only once in history (despite some honorable near-misses), and is easily shoved aside by inane but comforting superstitions. Even ordinary speech can be evasive.

Is it true, as Steven Pinker asserts, that both the reach and limitations of a language "organ," a human DNA-specified mental "instinct," power our speech and writing? The implications for the status of science as a special

D. Broderick, *Consciousness and Science Fiction*, Science and Fiction,
https://doi.org/10.1007/978-3-030-00599-3_2

way of knowing (like many playful science fiction thought devices, such as time machine paradoxes) are not self-evident, but they are worth teasing out.

Consider two bullets. One is simply dropped from near a gun's muzzle, the other fired horizontally, at the same instant. Which hits the ground first? Most of us have major trouble with this poser. Despite a lifetime of lifting and throwing and dancing, all our intuitions about motion start mumbling bad guesses. Most people conclude that the dropped bullet hits the deck first, probably much earlier.

Not so. Gravity pulls equally on both bullets. The sideways motion imparted by exploding gunpowder has no effect whatsoever on the rate at which the fired bullet falls to earth. How could it? (Well, it's true that the curvature of the planet does drop away but the extra vertical distance is negligible over the distance a bullet is likely to travel from a gun.)[1] But usually we need disciplined training in vector mathematics to understand this very elementary truth about how our world works. Science is not common sense. It is distinctly uncommon sense, and our brains—our minds—resist its enlightenment.

In recent decades, many sociologists have asserted that "western" science is just one form of many "ethno-sciences," each with its own rich claim to be taken seriously, and perhaps preferentially, as valid knowledge. We hear "wisdom traditions"—opinions formulated centuries or millennia prior to the formalized investigation of empirical reality—hailed as deeper and truer than evil reductive science. The late Alan Cromer, a particle theorist with a special interest in science education, denied this sort of apparently generous nostalgia. Most knowledge systems, he claimed, project the culture-bound shape of human minds upon the outside world. In a special sense defined decades ago by the developmental psychologist Jean Piaget, they are "egocentric."

By contrast, the techniques of inquiry invented by the Greeks and rediscovered 400 years ago in Europe—techniques which have remade our world utterly—deny that outer reality can be known through intuition alone. While the daily practice of science is clearly swayed by rhetorical skills, special interests and power politics, it works so well because at base it strives for empirical objectivity, reliable replication, and honest reporting. In Piaget's terms, its practice requires "formal operational" mental skills, which are never

---

[1] My rocket-science friend Spike Jones informs me: "If we assume a really high-velocity round, such as the AR-15. . . I find muzzle velocity of about 1000 meters per second, assuming the bullet is fired from about 2 m above the ground, takes about 0.6 s, so it goes about 600 m, difference in height due to earth's curvature is about 2.9 cm, which is small enough that we are splitting hairs if we argue (correctly) that the dropped bullet hits first, assuming no irregularities in the surface which of course we do have. If you could arrange a piece of ground which was perfectly planar with no irregularities, the two should hit about the same time, but aerodynamic effects could make it go either way. The difference in time is about 5 ms if we assume away the atmosphere, 5 ms being the time it takes for the bullet to travel the extra 3 cm due to curvature."

attained by more than half America's (and presumably all First World) adults. Hence, most of us "can't analyze a situation with several variables," as Cromer stated scathingly, "or understand a simple syllogism."

What of the intelligence of our evolutionary near-cousins, the apes? Steven Pinker has proposed Darwinian paths by which our mental modules evolved, including language itself. He is especially caustic about claims that bonobo chimpanzees have been taught a form of true sign speech. Pinker maintains that attempts to teach sign language or computerized lexical codes to chimps and other apes are doomed, because they did not share our eccentric evolutionary history. Human brains and other cultural organs, and the species-specific genes that dictate them, are precisely what make us volubly human and them merely clever apes.

But don't the chimps share 99% and apes 98% of their DNA with us?[2] True, but "the recipe for the embryological soufflé is so baroque," Pinker states, "that small genetic changes can have enormous effects." The contrary opinion was offered by primate specialist Sue Savage-Rumbaugh, who argued that Kanzi, a male bonobo, when young routinely combined symbols using "a primitive English word order [to convey] novel information." Pinker tried to show that Kanzi was just aping language. Such disagreements, of course, are the very stuff of a science that strives for objectivity even when it studies the mind. Certainly the evidence put forward by primatologists has regularly been assailed as selective, wishful at best.

Still, Savage-Rumbaugh and her late husband Duane Rumbaugh (who died in 2017) studied both common and bonobo chimps, notably Kanzi, for many years. (As I write, Kanzi is still alive at age 37.) In observations and experiments, they concluded that such primates can classify lexigrams (for example, into "food" vs. "tool" groups), zestfully play computer games using screen and joystick (as can rhesus macaque monkeys) and, arguably, employ grammatical constructions. Science fiction, as we shall see, has explored many alternative possible worlds where animals were "uplifted" or enhanced, and in which alien intelligences, with their own evolved distinctive communication systems, meet humans, usually to great confusion on both sides.

* * *

If the roots or templates of language are inherited, as they certainly are in humans and may be in bonobos, what of temperament and other apparently less rule-bound aspects of personality? Although the Greek physician Galen

[2] https://www.scientificamerican.com/article/tiny-genetic-differences-between-humans-and-other-primates-pervade-the-genome/

got a great deal wrong, his theories of physiology and anatomy ruled Western medicine for more than a thousand years. He is best known for his notion of the four humors—black and yellow bile, phlegm and blood—whose inherited dominance supposedly determined the patient's temperament: melancholic or choleric, phlegmatic or sanguine, or one of five mixed types.

The last living adherents of this quaintly old-fashioned doctrine were surely the Christian Brother teachers at my dismal trade school, but genomic temperament has been having an unexpected comeback. For decades, psychology taught that temperament, along with ability, was not innate but created by our experience of the world. Then some decades ago, with the discovery of neurotransmitters, we realized that brains and bodies are awash in chemical messengers, peptides, GABA, corticotropin releasing hormone and a hundred other tiny tides. The brain's logic circuits run as much on chemistry as on electric currents. And while our cortical and limbic circuits get tuned in an individual fashion, they tend to bunch into a few reliable categories, surprisingly close to Galen's humors. Research by Jerome Kagan, a Harvard developmental psychologist, mapped temperament and detected strong heritability. His team's results (spelled out in his 1994 book *Galen's Prophecy: Temperament in Human Nature*), for all their nuance and circumspection, inevitably stirred political passions.

There is good evidence that our feelings are organized in standard ways from earliest infancy, although styles of upbringing may skew the ways we act out those feelings. Some 15% of Caucasian babies studied were shy, timid or fearful, and these Kagan dubbed *inhibited*. They tended to have quite specific physical characteristics: narrow faces, pale eyes, tall, thin, allergic bodies. Brown-eyed, smiling, fearless, robust *uninhibited* children made up 30%. The rest fell somewhere between. (Chinese-American babies tended to be on average more inhibited.) Subtle new instruments and careful experiments endorsed this pattern, and helped explain the many acquired and hard-wired factors that build the profiles of temperament. Dominant activity in the left frontal brain relates to calm happiness, in the right with fear and sadness, while the right rear lobes govern emotional intensity, whether distressed or joyful.

Essentially, temperament seems to be a family of states organized around the level of reactivity of the sympathetic nervous system, which in turn is under the control of the amygdala, a part of the deep brain. High reactives are jumpy and reserved: worriers. Low reactives are extroverted, cheerful, fearless: warriors (or thugs).

* * *

But are "mere feelings" all that important? In the age of economic and other supposed rationalisms, is it not rather soppy and Sensitive New Age to care about emotions? (Discourse analysts have noted that such objections are coded sneers at traits historically regarded as "feminine" or "queer.") Actually, feelings are not only important to the quality of life but crucial to the human exercise of reason. As Antonio Damasio has made clear, feelings are "a window that opens directly onto a continuous updated image of the structure and state of our body."

René Descartes, you'll recall, supposed that we comprise a mechanical body yoked by divine contrivance to an impalpable and altogether finer self or soul, an immaterial essence designed to survive the corruption of the flesh. Since anyone could appreciate that pure minds can't get angry or horny or choked up with sentiment, bodily feelings had to be downgraded. Now we know differently. "The organism has reasons that reason must utilize," declares Damasio (in *Descartes' Error*), playing on Blaise Pascal's famous seventeenth century phrase, *The heart has its reasons, which reason knows not of.* Thoughts or ideas are "qualified" by feelings, which are markers within the body of how the world has affected us in the past. They are short-cuts to value: powerful devices that help guide us swiftly through the waffle of unchecked logic. Victims of pre-frontal leucotomy, whose links between the reasoning frontal lobes and the emotional amygdala have been cut, can lapse into a feckless inability to plan or decide. They can know but not feel. And so their knowledge is short-changed, their reasoning not merely "cold" but unhinged from reality.

* * *

Or is this analysis, in turn, a false polarity? Our culture's dearest tenet is the irrational as the source of the distinctively human, especially of creativity and madness. Our world, according to John McCrone, is shaped by the legacies of Romanticism, with its split between wild, unchecked forces of nature (emotion, good) and wussy restriction (reason, bad). When Freud told us the irrational Id (or It) had to be throttled and harnessed by the work-a-day Ego and Super-ego, we mourned the repressive cost of unbuttoned joy. Artists, lunatics and Byronic lovers elude these strictures, and Lacanian psychoanalysis attempted to track our own deepest impulses, rather mysteriously, to endlessly deferred "Desire for the Phallus" (which, unlike Freud's conjecture, has nothing much to do with the penis).

None of this is true, according to McCrone. We are made human by social language, as Russian linguist and philosopher Lev Vygotsky taught. Language completes our physical hardware (our bodies) with a specialized software (our minds). Your voice, literally, makes you human—the voice(s) we each speak within our own heads, telling ourselves the story of our social world. Feral children, lacking exposure to language in the first crucial years, never became properly human. Deaf mutes denied non-vocal coding languages such as Sign suffer tragically impaired identity.

Madness and dream result not from unleashing some deep irrationality or the slithering signifiers of Lacanian psychoanalysis, "brief glimpses of a higher reality" but (by and large) from "the broken-backed functioning of the bifold mind," as McCrone puts it (1993, p. 291). He means biological/hardware versus cultural/software elements. Thus, the confused tussle of an inner voice trying to make sense of a faulty biological ground of sensation—disrupted neurotransmitters, in the comparable cases of madness or drugs. In ordinary dreaming, stray vivid memory fragments are juggled in sleep by linguistic machinery optimized to narrate the tale of an organized external world.

Imagine, then, what dreams might plough the oceans of a brain scanned and replicated inside a machine (now a very familiar science fictional trope), a mind uploaded into a computer in order to escape the strictures of mortality. Would it find itself mad or newly reasonable? Sub-human in its extraordinary condition, or transhuman, even posthuman? Having a computer substrate would surely alter the experience of memory, which now shows us its treasures through such a narrow window and holds them in so vulnerable a storehouse. Even without uploading, of course, the memory of the technologically deathless might require substantial renovation and expansion, perhaps regular editing, or at least an improved means of filing and accessing its contents and links.

Although our organic human brains do not process memory in any manner as crass as "one neuron equals one memory" or even "one constellation of neurons equals one specific recollection," there are surely limits to how much we can cram inside our hundred billion or so brain cells and their webby synapses. One of the fruits of advanced neuroscience will be enhanced methods of storing and retrieving memories. For some of us, growing forgetful as we age, it won't come a day too soon.

Already, a blend of study skills and dedicated software agents and pharmaceuticals can improve effective memory and even the power of our thinking. Years ago a software program called Remembrance Agent or RA, developed by Bradley J. Rhodes and Thad Starner at MIT's famous Media Lab, could run in the background of your computer and keep tabs on everything that happened on your screen. The program compared each word it saw with a growing cross-referenced database derived from your email, the web sites you visit, other selected files and data services. A small window at the bottom of the screen ran a constantly changing reminder of these links, and if you wished to consult one of the listed files you just clicked on it. When you closed down your session's work, the RA updated its web of associations. As neuroscientist Anders Sandberg put it: "The RA combines the human, intuitive associative memory model with the efficient database memory of a computer." Used inventively, the RA and more recent programs of this kind can act as a cognitive accelerator as well as an aide memoire (in addition to Google and Wikipedia) to our grievously imperfect natural memory.[3]

The prospects of a posthuman consciousness are both exhilarating and alarming. Will enhanced humans even wish to conserve their original organic bodies? Consider this diverting anticipation by Russian-German Eugene Leitl (whose academic field is molecular modelling of wet, charged biopolymers such as peptides and lipids), a man not afraid to get right in your squeamish face:

> Apart from intrinsically higher and drastically cheaper diversity in artificial reality, a Dysonian computer cluster [where the planets of a star are reconstituted by an advanced science into a radiation-collecting shell with the surface area of a billion Earths] or a Jupiter brain is not a suitable environment for any classical body. Why limit oneself to human senses if I can utilize anything from proximal probes at the atomic scale to accelerometry, magnetometry, mass spectroscopy, particle detectors, huge interferometry arrays, etc? Exercise your own imagination on the motorics part. But, once again, if the material realm of a stellar system is entirely under your control, maxing out on computation while keeping the number of resources allocated to sensomotorics at an absolute minimum (in a pinch, you could always re-use these atoms in new configurations albeit at a time penalty and resources wasted in dormant basal auto-replicator capability) is obviously a good strategy. Only coevolution-artefacted nonlinearities would seem to disturb the Brave New Postmaterial world scheme.

Leitl (or 'gene, as he is known on the Internet) added with further wolfish humor: "I'd rather be a cluster of active-orbit-controlled boxes in high solar orbit swinging around my own gravitation center, while other parts of me are busily processing our stellar neighborhood. (Of course there will be probably no 'me,' just a memetic bouillabaisse virtually swirling in diverse bit-buckets of multiple shapes and sizes.)"

* * *

Yet, if some critics of orthodox AI are correct, even a brain the size of a gas-giant planet like Jupiter, or a Dyson sphere, could not be enough to sustain

[3] https://davidamerland.com/seo-tips/1033-google-is-the-remembrance-agent.html

a conscious mind running familiar computational programs. The famous British mathematician Sir Roger Penrose and his American colleague anesthesiologist Stuart Hameroff have long been exploring a quantum science of consciousness—or rather, trying to establish one. Mind cannot be understood and explained in crass computational terms, Penrose argues, without a new, extended kind of quantum physics that includes gravitation. This would incorporate the realm of relativity, Einstein's powerful mathematical framework of spacetime, which to date has not been integrated satisfactorily with the quantum realm of quarks, electrons and light. Penrose is a Platonist: like Max Tegmark, he believes in a mathematical reality deeper and more primordial than the world we observe.

In several books dense with quantum theory and other tricky mathematics (allegedly simplified for the ordinary reader, but painfully difficult nonetheless), he has tried to convince us that human brains can do things that computers never will. If that's true, what is it about our mind/brains that makes us special? Do we, after all, have immaterial souls able to leap free of the restrictions that apply to matter? No, says Penrose, but we do have brains that might utilize mysterious quantum-realm abilities.

His key is Hameroff's theory, based on the allegedly strange properties of neural cytoskeletal microtubules, tiny hollow nano-scale tubes inside cells built from columns of tubuline dimers (protein polymers) able to switch between two conformations. Most physiologists see these microtubules as nothing more exciting than structural struts holding the cell nicely in shape. The quantum-mind theory speculates that their signal-transmitting properties might permit parts of the brain to act as cellular automata, or even non-locally (that is, contrary to the ordinary limits of space, time and causality), and thereby surpass their crass physical limitations.

Hameroff presents the bones of this case:

> What consciousness actually *is* and how it comes about remain unknown. The general assumption in modern science and philosophy—the "standard model"—is that consciousness emerges from complex computation among brain neurons, computation whose currency is seen as neuronal firings ("“spikes") and synaptic transmissions, equated with binary "bits" in digital computing. Consciousness is presumed to "emerge" from complex neuronal computation, and to have arisen during biological evolution as an adaptation of living systems, extrinsic to the makeup of the universe. On the other hand, spiritual and contemplative traditions, and some scientists and philosophers consider consciousness to be intrinsic, "woven into the fabric of the universe." In these views, conscious precursors

and Platonic forms preceded biology, existing all along in the fine scale structure of reality.

My research involves a theory of consciousness which can bridge these two approaches... it suggests consciousness arises from quantum vibrations... which interfere, "collapse" and resonate across scale, control neuronal firings, generate consciousness, and connect ultimately to "deeper order'" ripples in spacetime geometry. Consciousness is more like music than computation.[4]

It is a subtle and perhaps romantic argument. Few specialists in any of the associated fields, however, are prepared to follow the speculations of Penrose and Hameroff (although they are attended to with a measure of respect). Interestingly, Penrose remains a reductionist, believing that one day we will possess theories robust enough to let us understand even the incalculable mind. I feel fairly sure that he is wrong in claiming that minds cannot be emulated by conventional but massively enhanced non-quantum computers. (In Chap. 11 we shall look at a science fictional version of Penrose's model in Robert J. Sawyer's *Quantum Night*.)

Meanwhile, Daniel Dennett's approach, rich with detail from brain science, argues that most living creatures lack consciousness but are no worse off for this: they possess *competence without comprehension.* With buoyant T.H. Huxleyite clarity, he tracks four stages of evolution (*From Bacteria to Bach and Back*, 2017, pp. 98–9). First there was the simple *Darwinian* branch of life, still healthy and vastly numerous; entities "born 'knowing' all they will ever 'know'; they are gifted but not learners." Next came *Skinnerian creatures*, shaped beyond their inherited evolutionary dispositions by teachable moments of reinforcement, the basis of operant conditioning. Then arrived *Popperian creatures*, able to form primitive conjectures and test them so "their hypotheses die in their stead" (as Karl Popper put it). Finally, we arrive at *Gregorian creatures*—named not for Popes, but for psychologist Richard Gregory. These creatures deliberately introduce thinking tools and teach their young the use of them, and perhaps how to go beyond them: "arithmetic and democracy and double-blind studies..." So far, *Homo sapiens* is the only certified Gregorian species. Artificial General Intelligence might be the next to emerge.

Let us propose, then, that consciousness is built upward or outward from the senses (and perhaps conceptual templates) we inherit, those shaped by the capacities they afforded in the environments our ancestors enjoyed or suffered. Consciousness, then, is the state or condition of being aware of the world (and of the self, however torn or fragmented or task-dedicated) via the evidence

[4] https://www.scienceandnonduality.com/speakers/stuart-hameroff/

from the senses and the degree of successful match of our composite portraits of these experiences, and our memories of them and, from those, the projections of what we can expect in the future.

Even if no current approach is correct, mind will continue to be the proper, if baffling, study of an enhanced humanity—and of the science fiction that here and now projects multiple paths into those baffling and exciting futures.

3

What Is It Like to Be a Conscious SF Writer?

If consciousness is distributed across billions of individual neurons that are located in innumerable brain regions at different hierarchical levels of the nervous system, does it follow that each of these neurons individually "possesses" consciousness? And even if these neurons are networked together, is there something physically unified in the brain that has the same grain as the unified mind?
Todd E. Feinberg. M.D., *2001, p. 117*

In 1974, the philosopher Thomas Nagel made his mark with a pungent little challenge to his colleagues who agreed that "Consciousness is what makes the mind–body problem really intractable" (Nagel, p. 391). "[W]e have at present no conception of what an explanation of the physical nature of a mental phenomenon would be" (pp. 391–92). Here was his suggested pry-bar, somewhat like a *koan*, in the title of his paper: "What Is It Like to Be a Bat?"[1]

Fundamentally, he claimed, an organism has conscious mental states if and only if there is something that it is like to *be* that organism—something it is like *for* the organism (p. 392). That is, he was probing "the subjective character of experience." And Nagel sought for something beyond imaginative emulation of life as a nearly blind insectivore with webbed arms for gliding and echolocation as a surrogate for visual mapping of the world. That would tell him "only what it would be like for *me* to behave as a bat behaves. But… I want to know what it is like for a *bat* to be a bat" (p. 394).

[1] Nagel, in Hofstadter and Dennett, *The Mind's I* (Penguin 1982).

D. Broderick, *Consciousness and Science Fiction*, Science and Fiction,
https://doi.org/10.1007/978-3-030-00599-3_3

In short, what Nagel sought was the specifics of subjective experience in any given organism, in order to account for its consciousness. For a human to be *un*conscious was precisely to lack (temporarily, we might hope) the awareness of being a self, situated in space and time, probably a member of a group of other humans, and so on. In a sense, this challenge might seem circular and uninformative: *what does it feel like to feel?* But let us press this query further, into the space of science fictional thought experiments: *What is it like to be a Martian*?

Well, Nagel is there ahead of us. If there's "conscious life elsewhere in the universe, it is likely that some of it will not be describable" by us. We are in a similar position to that which "Martians would occupy" *vis-à-vis* conceiving what it's like to be a human. But while human language will probably never be able to accommodate "a detailed description of Martian or bat phenomenology," we have no warrant to dismiss, without sharing it, the richness of their sui generis experiences.

A significant portion of science fiction attempts the impossible task of rendering vividly the consciousness not just of unusual, mutant or bioengineered humans, but also of enhanced animals from earthly stock and alien life forms ranging from swarms of quantum-scale intelligences to aware stars and entire galaxies. Indeed, since before science stood on the brink of creating intelligent machines equal in mental capacity to humans, sf had been exploring the imagined consequences of robots (often these days known as "bots"), and the question whether they must necessarily be mindless zombies or as conscious as waking humans. In Nagel's sense, we must ask *What is it like to be a bot?*

The famous science fiction name that instantly springs to mind (for those with minds) is "Isaac Asimov," who was not the first writer to envisage conscious machines but put his stamp on the concept forever with his Three Laws (later Four) of Robotics. These were designed to safeguard the bots' creators as well as the robots themselves, and arose in computer programming written into their "positronic brain" networks. Robots were clearly aware of these constraints, since breaching these laws led swiftly to brain-locking inanition, usually unrepairable: robot death.

In the same vein, sf explored artificial intelligence residing in fixed, non-anthropomorphic computers, as well as enhanced composites, organic humans boosted and extended by mechanistic amplification. We shall look at a selection of such not quite robotic robots as portrayed with science fiction, from early guesses to current informed projections.

* * *

But still the core question "What is consciousness?" troubles many people. As noted briefly in our first chapter, proffered answers range from the banal and underpowered to the apparently ludicrous. Most of these alleged insights—some emerging from neuroscience and brain surgery, some from philosophical and theological musing—have worked their way into science fiction stories, novels, cartoons, computer games, and TV or movies. The most direct explanations reflect either physicalist (or "materialist") theories or their idealist (perhaps "immaterialist") alternatives, with Cartesian mind–body dualists trying to solder the two together into a working partnership.

On the strict reductionist account, sometimes dubbed "eliminative materialism," there is no Hard Problem of consciousness. It is just the fabulously elaborate nervous system doing its job of keeping the body ticking over, surveying the local environment, keeping tabs on likely opportunities ("affordances") and hazards. As John Searle noted:

> By "consciousness" I simply mean those subjective states of sentience or awareness that begin when one awakes in the morning from a dreamless sleep and continue throughout the day until one goes to sleep at night, or falls into a coma, or dies, or otherwise becomes, as one would say, "unconscious."
>
> Above all, consciousness is a biological phenomenon. We should think of consciousness as part of our ordinary biological history, along with digestion, growth, mitosis and meiosis. (*Consciousness and Language*, p. 7)

If so, there is no need for a soul or embodied entity meant for better prospects after death and its release from embodiment. Notable theorists of this option include long-time philosophy professors at the University of California, San Diego, Patricia and Paul Churchland, and perhaps Daniel Dennett at Tufts, notable for his bulky 1991 book *Consciousness Explained.* As Dennett wryly complains (in a later volume, *Intuition Pumps*, 2013), many readers were inspired "to joke that my book should have been entitled *Consciousness Explained Away* or... *Consciousness Denied*" (p. 313).

Actually this was rarely a light-hearted *joke*; more often, it was a denunciation. Patricia Churchland responds to this kind of canard with some asperity in *Touching a Nerve* (2013): "I entered an elevator, joining an anthropologist who had got on earlier. My very presence brought her to fury, and she hissed: 'You reductionist! How can you think there is nothing but atoms?' Who, me? I was flabbergasted" (p. 262). Here is Churchland's defense:

> Reductionism is sometimes equated with *go-away-ism*—with claiming that some high-level phenomenon does not really exist... If, as seems increasingly likely, dreaming, learning, remembering, and being consciously aware are activities of the physical brain, it does not follow that they are not real. Rather, the point is that their reality depends on a neural reality. If reduction is essentially about explanation, the lament and the lashing out are missing the point. (pp. 262–63)

Curiously, one of their sharpest-tongued critics is the Catholic Thomist philosopher and vehement Aristotelian Edward Feser, in no way to be confused with the atheistic and equally vehement Aristotelian Ayn Rand. In his *The Last Superstition: A Refutation of the New Atheism* (2008), Feser insists that eliminative materialism is a dead end, the poisoned result of Western thinkers having abandoned Aristotle's theory of *formal* and *final* causes. The former Hylomorphic doctrine sees *being* as a compound of crude matter and immaterial form (perhaps a bit like hardware and software, or the conjoining of canvas and oils with a painter's mental intentions to form a work of art). The latter is the pre-Darwinian proposition that each event is predestined, directed in advance to a purposed end: that is, it is teleological. "[T]o say that matter, understood in mechanistic terms, is *all* that exists, is implicitly but necessarily to deny that the mind exists" (pp. 236–37).

There is one unfashionable notion toward which this assertion leads (*pace*, Ms. Rand), and Feser states it baldly:

> the entire conception of God enshrined in classical monotheism, the immortality of the soul, and the natural law system of morality... In fact, the material world points beyond itself to God; but the secularist sees only the material world. The material side of human nature points beyond itself to an immaterial and immortal soul; the secularist sees only the material world. (p. 267)

For Dennett, what the material and informational world points to is the community of individual communicators: "consciousness is not just talking to yourself;" (or, one might add, to an imaginary deity or several of them) "it includes all the varieties of self-stimulation" (no coarse jokes, please) "and reflection we have acquired and honed throughout our waking lives. These are not just things that happen to our brains; they are behaviors we engage in ... some 'instinctively' (thanks to genetic evolution) and the rest acquired (thanks to cultural evolution and transmission and individual self-exploration" (*From Bacteria to Bach*, p. 346).

Yes, but—how does any of this elaborate process end up with feelings and outpourings of emotion and detestation of drudgery, rather than passionless cortical or thalamic meter readings? What is it like to be something that it likes to be?

* * *

Dismissing any doomed attempt to revive Aristotle's metaphysics, today's immaterialists and dualists draw attention to the persistent mystery of consciousness as an alleged side effect of three pounds of fatty and proteinaceous tissue under the skull. Why and how does this brute if complex machinery of cognitive and emotional systems give rise to *experiences* rich in subjective *feelings*? What can be the naturalistic, non-religious origin of minds capable of reflection and nuanced planning far exceeding the automated algorithms governing the activity of plants and non-human animals?

As we have been reminded several times already, a widespread and ancient answer tells us that the fleshy brain (or heart, or belly, according to cultural taste) is coupled to a kind of ghost or spirit with special abilities luckily suited to the job of turning neuronal sense impressions into *qualia*, "the introspectively accessible, phenomenal aspects of our mental lives."[2] Just how something with no observable linkage to spacetime and matter–energy can do this remains an acknowledged puzzle. To bypass this difficulty, one of the increasingly popular theories developed by these analysts is that consciousness is itself a basic component of everything in the universe, comparable to spacetime itself, and energy and matter (which itself can be regarded as a kind of congealed energy).

Perhaps not surprisingly, Professor Nagel is a leading exponent of this viewpoint. Though an atheist with no fondness for ghosts or spirits, he supports the doctrine known as *panpsychism*—mind literally everywhere, in differing grades of complexity. Consciousness, on this account, is rather like mass or inertia, just an attribute of every entity we observe (including, of course, ourselves), all the way down. Somehow, when the primitive consciousness elements are arranged in just the right way, they form an informational system that becomes, in varying measure, self-aware. An increasingly popular neuroscience hypothetical model of this architecture is University of Wisconsin's Professor Giulio Tononi's Integrated Information Theory (IIT). This maps degree of consciousness, dubbed Phi, by strength of neural connectivity and mutual activation. In human beings—and to some degree, presumably, some of the savvier animals—this patterned ensemble gives rise to all the mysterious phenomena of mental life: brilliant hues and wafting scents or vomitous stinks, symbolic gestures, utterances and inscriptions, irritability and devoted love, dreams (or wait, are they by definition *un*conscious?), memory and repression, hopes, plans, regrets, urgently embraced beliefs... But how does it *do* this? The answer is not yet forthcoming, despite the efforts of Tononi and his colleagues:

> The science of consciousness has made great strides by focusing on the behavioural and neuronal correlates of experience. However, while such correlates are important for progress to occur, they are not enough if we are to

[2] https://plato.stanford.edu/entries/qualia/

> understand even basic facts, for example, why the cerebral cortex gives rise to consciousness but the cerebellum does not, though it has even more neurons and appears to be just as complicated.[3]

The apogee, or nadir, of the immaterialist preference is the assertion that consciousness is not just an added gadget in the menagerie of particles and force fields, equal to them but providing its own special contributions (qualia, and so on). No, it is metaphysically *prior* to everything else, universe/s included. Does that mean it is God, or a community of creative and sustaining gods? Just when canonical science, and most science fiction, thought it had finally struggled free of theology's brambles, are its theoreticians obliged by sheer logic and deduction to reinsert Deity into the recipe of Being and Consciousness?

Perhaps surprisingly, this counterintuitive theodicy does crop up now and then in science fiction, sometimes under the influence of quasi-Eastern mystical doctrines such as Madam Blavatsky's Theosophy, sometimes directly from Indian religious teachings, in particular Hindu proposals such as Universal Consciousness/Absolute Consciousness (Brahman, Atman). My Indian colleague Dr. Sonali Mawaha points to a frequent blurring of any distinction between experience and experiencer:

> Adi Shankracharya's Advaita Vedanta clearly lays out this view of the Paramarthika satya (Absolute Truth), Self, and self, as the ultimate idealist view, considering everything as a projection of the mind, an illusion (Maya). However, he was forced to account for the truth of the lived reality (vyavahrika satya) to account for the basic survival needs (food, for instance). The Advaita Vedanta is just one of several interpretations of the Vedanta.[4]

An extremely simplified version of these doctrines formed the basis of pulp sf writer L. Ron Hubbard's Dianetics and Scientology in the 1950s and beyond. Hubbard was a charismatic fraud who taught his devotees that the phenomenal world is constructed by the mind, and can be readily manipulated by consciousness once the invading nonmaterial and extragalactic thetans are put in their place (or something; the dogma kept changing, and prices went up as the gullible sought higher understanding and command of their own intergalactic being). He had been for some years a close friend of Robert Heinlein, who was arguably the founder of what has come to be called "Modern science fiction." Heinlein's own fiction, as we shall see, sometimes

[3] Giulio Tononi & Christof Koch, 2015.

[4] Here is a brief summary: http://vedantastudent.blogspot.in/p/essence-of-advaita-vedanta.html

made use of this kind of ontology, which also crops up now and then in work by writers influenced by him.

A minor, sentimental early example is Heinlein's novella "Waldo" (1940). From infancy, the genius protagonist Waldo Farthingwaite-Jones has been afflicted with *myasthenia gravis*, an autoimmune disease causing muscular enfeeblement so severe that he had to spend his life in "a condition of exhausted collapse" (p. 14). Despite this desperate drawback, Waldo makes a great deal of money as an inventive engineer of telefactors later known as "Waldos": artificial limbs and hands that mimic his own weak movements and magnify them, financing relocation to a luxurious geostationary space station. Here, in null-gravity, the pain and indignities of his condition are minimized, but not his bitter sourness, while fresh inventions pour from his workshops on Earth.

Meanwhile, broadcast electric power, along the lines sought by Tesla, makes oil, coal and nuclear sources redundant—until suddenly it stops working. Planes and cars crash, factories grind to a halt. Waldo finds, against his will, that the power of the deKalb receptors is driven by something uncanny, to be mastered by backcountry hexes. "Magic is loosed in the world!" Returned painfully to Earth, Waldo learns to reach into "the Other World" and draw energy from its mystic provenance, allowing him to abolish broadcast power and the subtle radiation damage its use has inflicted on the world's populations. Waldo himself is healed, and his genius flowers into neurosurgery and fabulously joyous tap dancing. Now everyone admires and loves Waldo. Consciousness, it might be said, has reached into secret places and unleashed forces suspected previously, and sometimes put into effect, only by the superstitious.

* * *

More than half a century later, such ideas are re-emerging alongside claims that "Consciousness is a Fundamental," as writer and mind explorer Stephan Schwartz puts it, finding support in a number of somewhat obscure statements by European quantum theory pioneers from a century ago: Niels Bohr especially, Max Planck, Wolfgang Pauli, Werner Heisenberg, Erwin Schrödinger but not Paul Dirac, Einstein and others. Schwartz frequently cites a statement from Planck (1858–1947; Nobel laureate in Physics 1918):

> This work is pushing toward a new paradigm, one that is neither dualist nor monist, but rather one that postulates consciousness as the fundamental basis of reality. Max Planck, the father of Quantum Mechanics, framed it very clearly in an interview with the respected British newspaper, *The Observer* in its January 25, 1931 edition. Context is always important, and Planck understood very well that he was taking a public position, speaking as one of the leading physicists of

> his generation, through one of Britain's most important papers. He did not mince words: "I regard consciousness as fundamental. I regard matter as derivative from consciousness. We cannot get behind consciousness. Everything that we talk about, everything that we regard as existing, postulates consciousness." (Stephan A. Schwartz, 2015)

This idea forms the spine of Schwartz's recent science fiction novel *The Awakening: A Novel of Aliens and Consciousness* (2017), which we shall consider in more detail in Chapter 11. A crashed UFO alien, captured by US military and intelligence agencies, reveals itself to several humans to whom it teaches the mysterious wisdom and science 10,000 years in advance of our species' own. These are revealed to us in extended passages echoing the author's own proclamations on "nonlocal consciousness" and "information architectures."[5]

Once an influential research paradigm, Niels Bohr's "complementarity" was patently inflected by his era's culture, captured by Harvard historian Juan Miguel Marin in a 2009 paper published in the *European Journal of Physics*, and summarized below by Phys.org's Lisa Zyga:

> [Q]uantum mechanics up to World War II existed in a predominantly German context, and this culture helped to form the mystical zeitgeist of the time. The controversy died in the second half of the century, when the physics culture switched to Anglo-American. Most contemporary physicists are, like Einstein, realists, and do not believe that consciousness has a role in quantum theory. The dominant modern view is that an observation does not cause an atom to exist in the observed position, but that the observer finds the location of that atom.
>
> ...Schrödinger's lectures mark the last of a generation that lived with the mysticism controversy. (Lisa Zyga, 2009)

A more searching and developed survey of this topic is astrophysicist Adam Becker's detailed popular treatment *What Is Real? The Unfinished Quest for the Meaning of Quantum Physics* (2018), which assails Bohr's all but impenetrably congested prose and that of his once dominant Copenhagen model, championing more recent approaches to the puzzles of quantum theory. This is clearly fertile territory for informed science fiction writers and readers, and will continue to be so until the nature of reality is finally nailed down to the satisfaction of science and latter-day mystics alike.

* * *

[5] For example, https://www.explorejournal.com/article/S1550-8307(16)30115-X/fulltext

Still, as noted, mystical views of consciousness do seem to be once again on the rise. A similar cryptic credo is presented in an amusing way as the Preface to Dr. Dean Radin's *Real Magic* (2018), a study in science and the paranormal. Regrettably, that book bears the disquietingly catchpenny subtitle *Ancient Wisdom, Modern Science, and a Guide to the Secret Power of the Universe.* Here we read, as if in the opening of a science fantasy book, an imaginary Guest Editorial from the future New Seattle Province, June 1, 2915. (Polymath Max Tegmark uses a similar narrative device to open *Life 3.0*, his 2017 exploration of a near-future Artificial General Intelligence. His "Tale of the Omega Team" presents a brief fanciful history of an AGI's creation and flowering that brings about a kind of planetary utopia of abundance and freedom.)

In Radin's version, a fragment of a centuries-old computer file has been located and rendered readable; this is the lost text of an editorial from the long defunct *Galactica Today*. Populations had embraced demagogues and resentful tribalisms as global warming and plagues afflicted the world. Luckily, it was realized at last that all these grim assaults were "rooted in humanity's faulty understanding of consciousness, which, as we now know, is the fundamental glue that binds the fabric of reality" (*Real Magic*, p. ix).

Matters improved at the end of the twenty-first century with the full realization that mind can directly shape external-reality,

> when Hilda Ramirez of Human University first conclusively demonstrated the plasticity of physical reality. Her evidence that the speed of light and other physical constants were mental constructs, not inviolable absolutes, provided a clear path to global harmony. (p. x)

How could such knowledge have been overlooked for centuries, decried as wicked demonic "magic" or fake sorcery? Well, humanity had suffered during our own bleak times:

> . . .the most educated minds had convinced themselves, despite an enormous body of evidence to the contrary, that reality emerged solely from various forms of energy. Their crude instruments were unable to detect the multidimensional tapestry of consciousness. (p. xi)

This is an entertaining conceit that could form the background to a series of sf novels, and indeed already had in Frank Herbert's *Dune* sequence 50 years previously. Taken as the literal truth, is it possible for serious scientists (I am not thinking of Deepak Chopra, for depressing example) to embrace it? To some extent, yes. Philosophy professor David Chalmers might be slipping from a panpsychist solution to his Hard Problem of consciousness all the way into a revival of metaphysical idealism. His essay "Idealism and the Mind–Body Problem" opens by quoting a sharp, perhaps cynical witticism:

> "One starts as a materialist, then one becomes a dualist, then a panpsychist, and one ends up as an idealist".... First, one is impressed by the successes of science, endorsing materialism about everything and so about the mind. Second, one is moved by the problem of consciousness to see a gap between physics and consciousness, thereby endorsing dualism, where both matter and consciousness are fundamental. Third, one is moved by the inscrutability of matter to realize that science reveals at most the structure of matter and not its underlying nature, and to speculate that this nature may involve consciousness, thereby endorsing panpsychism. Fourth, one comes to think that there is little reason to believe in anything beyond consciousness and that the physical world is wholly constituted by consciousness, thereby endorsing idealism. (2018)

Would this be a sort of quantum Kantianism or Hegelianism? Not for Chalmers, who notes candidly:

> Absolute idealism is typically associated with a number of Hegelian doctrines concerning teleology and rationality, and I do not have a clear sense of how these doctrines bear on the mind–body issues I am concerned with here. The label is occasionally used more straightforwardly for an idealism grounded in the mental states of a single cosmic entity, but to avoid the resonant Hegelian overtones I will give that view a different label... cosmoidealism...

One clear objection—although it was not really intended as such—is Max Tegmark's physics-based rejection of any Cosmic Mind because of its global mental sluggishness. The more extensive the mind, the slower its global consciousness runs to allow all its parts to share information. Tegmark calculates that because of communication time lags an immense mind the size of a galaxy "could have only one global thought every 100,000 years or so... no more than a hundred experiences during the entire history of our Universe thus far!" (*Life 3.0*, p. 309). (This appears to be an error of simple arithmetic: since the cosmos is less than 14 billion years old, there would be roughly 140,000 consecutive instances of a hundred thousand years. On the other hand, in the beginning of time the universe was still very compacted, so if a Cosmic Mind was around back then its integrated information would race around the cosmos very much faster.)

Since usable communication of information is limited by the relativistic bound of the speed of light, this means that the more grandiose and vast the Mind of the Universe becomes as spacetime keeps expanding, the slower it will function. Its component parts (assuming It has anything so metaphysically lowbrow) might contain truly immense amounts of data and analyses, but It will not be able to access and share these thoughts and memories beyond a fairly defined limit without slowing to the point where even a short string of binary digits will take millions or billions of years to traverse Its expanse.

Of course, those who embrace cosmoidealism without quite rejecting standard physics might clutch at nonlocal entanglement as a quantum solution—but this real phenomenon, according to everything science knows at the moment, is incapable of transporting information from one place to another in a form that can be acted on "faster than light." Maybe very thin and robust hyperspatial wormholes might serve as instantaneous conduits, carrying information through higher dimensions, like a string stretched tight to connect two tin cans. But for physics, at this stage, that really remains a can of worms. (Sorry.)

Chalmers concludes his own discussion: "I think cosmic idealism is the most promising version of idealism, and is about as promising as any version of panpsychism. It should be on the list of the handful of promising approaches to the mind–body problem," but he adds, frankly, "I do not claim that idealism is plausible. No position on the mind–body problem is plausible."

In any event, it will be interesting to see how any such a shift to an idealist explanation for consciousness, if it proves to be persuasive, influences science fiction in coming years and decades.

4

The First Century-Plus

The idea that complex, abstract abilities, or talent, can somehow be found in the linear code of DNA, and with it the notion that genes act as such independent units, is as superstitious as the idea that they can be implanted by passive experience of instruction.
Ken Richardson, *1999, p. 237*

Can any century really be nominated as the founding moment of *fantastika* in its science fictional avatar? Many scholars are eager to press back all the way to the composition of *The Epic of Gilgamesh* (which came together around 4000 years ago) or at least pioneers such as Lucian, with his *True History*, Ovid with *Metamorphoses* and even earlier mythological quasi-sf such as the Hindu *Ramayana* and *Mahabharata*.

But the sf mode or genre did not truly arrive (according to the magisterial online *Encyclopedia of Science Fiction*) with a "sense of the possibilities of change," especially driven by technology, until the seventeenth century nor "percolate into society at large" until the eighteenth and nineteenth centuries, with emphasis on the latter.[1] In the early nineteenth century it emerged blinking and moving its damp wings cautiously in Gothic tales by Mary Wollstonecraft Shelley and Edgar Allan Poe. It formed the basis for adventures by Jules Verne in the middle of the century, and attained its first real identity and maturity in the brilliant, innovative very late nineteenth century "scientific

[1] http://www.sf-encyclopedia.com/entry/history_of_sf

D. Broderick, *Consciousness and Science Fiction*, Science and Fiction,
https://doi.org/10.1007/978-3-030-00599-3_4

romances" by H.G. Wells. Let us examine some of these early experiments in new narrative, before moving into the thick of the following century, and more, of commercial sf.

* * *

What Is It Like to Be a Patchwork Monster?

1818/1831 Mary W. Shelley, *Frankenstein, or, The Modern Prometheus*

Why "Prometheus"? Because that Titan, in ancient Greek mythology and prior to the Olympian gods, gave the first humans the secret of fire, and suffered endless agonies for this apostasy. The young eighteenth century chemist Victor Frankenstein, of Geneva in Switzerland, emulates that crime by constructing a human-like body from rotting bones and flesh and uses his unexplained discovery to bring the patchwork composite to life.

This secret is not, despite its typical portrayal in movies, a terrifying flash of captured lightning that slams electricity into imbedded electrodes (not *bolts* in Boris Karloff's neck, an icon of crude mechanism, as many believe.) Appalled less by his impiety than by the sheer ugliness of this eight-feet tall, broad-shouldered neonate, Victor quivers for a time in his bedroom and then rushes back and forth in the street. When he returns, the creature is gone.

Curiously, the nameless composite being shows no evidence of memory, although he clearly possesses a recycled brain—possibly the finest the world has ever known. We watch him teach himself first single lexemes and fairly quickly elegant French or German, gained by peering through a crack in a hovel wall and listening to a few people going about their daily business. Here is a consciousness apparently fresh-hatched and entirely self-taught, under the grimmest restrictions, with no mother (like Mary Shelley herself, whose notable mother Mary Wollstonecraft died when her daughter was a month old) and disowned by his male creator. So while the creature has no explicit recollection of a previous life, perhaps we might conjecture that his brain retains the imprint of grammatical linguistic competence (proof of the theories of Noam Chomsky and Steven Pinker).

In any event, this unattractive monster begins confused but benign, hungry for companionship but meeting only screams of terror at his brutish appearance, beaten when he tries to help a child in peril. As the monster recounts,

> My heart was fashioned to be susceptible of love and sympathy; and, when wrenched by misery to vice and hatred, it did not endure the violence of the change, without torture such as you cannot even imagine... When I first sought [sympathy], it was the love of virtue, the feelings of happiness and affection with which my whole being overflowed, that I wished to be participated. But now, that virtue has become to me a shadow...

A ship's captain near the Arctic ice finds Dr. Frankenstein years later marooned on the ice, having pursued the immense "daemon" sighted briefly by the crew. Victor's appearance is not unlike his creature's at its inception. The captain writes:

> I never saw a more interesting creature: his eyes have generally an expression of wildness, and even madness; but there are moments when, if any one performs an act of kindness towards him, or does him any the most trifling service, his whole countenance is lighted up, as it were, with a beam of benevolence and sweetness that I never saw equaled. But he is generally melancholy and despairing; and sometimes he gnashes his teeth, as if impatient of the weight of woes that oppresses him.

Unlike the abandoned and despised creature of Frankenstein's scientific paternity, though, he is swiftly embraced by these fellow humans:

> ...his manners are so conciliating and gentle, that the sailors are all interested in him, although they have had very little communication with him. For my own part, I begin to love him as a brother; and his constant and deep grief fills me with sympathy and compassion. He must have been a noble creature in his better days, being even now in wreck so attractive and amiable.

Several hundred pages of correspondence and transcribed histories further on, including a tragic testament from the unfortunate "monster," Victor prepares for death, addressing Captain Walton, his nautical rescuer:

> Seek happiness in tranquility, and avoid ambition, even if it be only the apparently innocent one of distinguishing yourself in science and discoveries.

As he dies, Walton is plunged into grief:

> he pressed my hand feebly, and his eyes closed for ever, while the irradiation of a gentle smile passed away from his lips. . . .what comment can I make on the untimely extinction of this glorious spirit? What can I say, that will enable you to understand the depth of my sorrow? All that I should express would be inadequate and feeble. My tears flow; my mind is overshadowed by a cloud of disappointment.

The creature appears in the cabin, aghast at the death of his parent and his own crimes of vengeance. Walton sees him at last:

> gigantic in stature, yet uncouth and distorted in its proportions. As he hung over the coffin, his face was concealed by long locks of ragged hair; but one vast hand was extended, in colour and apparent texture like that of a mummy. When he heard the sound of my approach, he ceased to utter exclamations of grief and horror, and sprung towards the window. Never did I behold a vision so horrible as his face, of such loathsome, yet appalling hideousness. . . I dared not again raise my eyes to his face, there was something so scaring and unearthly in his ugliness.

Much hangs on how physical beauty versus ugliness is presented as a clear and unquestioned index of the several characters' moral virtue. From the first moment Victor sees his botched creation come alive, he calls him a fiend (a term, as we shall see, applied also to Mr. Hyde in Robert Louis Stevenson's novella of drug-induced dissociative personality disorder nearly seven decades later):

> . . .by the glimmer of the half-extinguished light, I saw the dull yellow eye of the creature open; it breathed hard, and a convulsive motion agitated its limbs.
>
> How can I describe my emotions at this catastrophe, or how delineate the wretch whom with such infinite pains and care I had endeavoured to form? His limbs were in proportion, and I had selected his features as beautiful. Beautiful!—Great God! His yellow skin scarcely covered the work of muscles and arteries beneath; his hair was of a lustrous black, and flowing; his teeth of a pearly whiteness; but these luxuriances only formed a more horrid contrast with his watery eyes, that seemed almost of the same colour as the dun white sockets in which they were set, his shrivelled complexion and straight black lips.

In the twenty-first century, no doubt we witness this projection of wickedness upon mere physical deformity with disgust and disapproval, even though we might be equally terrified at the sight and smell of such a being, especially in view of his known murders and vengeful cruelty. But what is noticeable in the passage from creation to escape is the being's initial innocence and optimism; the monster is a kind of saint seeking only to share love. This is apparent in his demand that Victor build for him a monster-wife, so that even in the wilderness he need not suffer loneliness and the inability to share his own undeniable if sui generis consciousness with another.

Victor's own tormented and tormenting consciousness is displayed in detail, and we witness his obsessive ambition, his unreasoned revulsion against his (as it were) clone, or child. Scholars have noted Mary Shelley's own suffering as first her mother perished, then her own children. She was pregnant five times, and all but one died soon. When she was 18 her premature daughter died, less than a year later she had a son, William, who died from malaria at the age of two and a half, there were suicides in the family, her daughter Clara Everina died in 1818. It was an era lacking antibiotics and racked by fatal illnesses. Only her son Percy Florence survived (and for 70 years). Her beloved husband Percy Bysshe Shelley drowned at sea when Mary was not yet 25 and he was nearly 30; his previous wife had drowned earlier. It seems clear that this somber history inflected at least the final, extended release of *Frankenstein.*

Was the novel, as argued by Brian W. Aldiss, really the first genuine work of science fiction? Despite its moments of gaudy murder, revenge, flight, all in a Gothic mode, Aldiss makes a strong case and one that has been widely adopted. It is perhaps startling that Samuel R. Delany referred in a 2018 Facebook post to "Brian Aldiss's lunatic 1973 notion that Frankenstein has anything to do with science fiction's origins." This extravagant dismissal is unwarranted. Other evaluations are equally odd. In 2007, a gay academic, John Lauritsen, published *The Man Who Wrote Frankenstein*, claiming that Percy Bysshe Shelly was obviously the true author of a book Lauritsen regarded as a poem to male mutual love. Certainly this (closeted) aspect will strike today's readers as plausible in a degree unimaginable half a century ago or longer.

Germaine Greer rejected this claim of pseudonymous authorship as absurd, not on the expected grounds that a woman like young Mary could certainly write a great novel but rather because Greer found the book shoddy and poorly constructed. This, too, is true in some degree, if it is read as entirely naturalistic realism. For instance, it is mysterious how Victor can build an eight-foot tall male with broad shoulders out of available corpses, let along do this work sans refrigeration in his attic flat, without fellow residents or neighbors noticing the stench, even in the rancid eighteenth century. There is a passing mention of a servant, but none of huge amounts of ice being carried up the stairs and bloody water and rotting tissues carried down again for disposal. (Given the emphasis on icy mountain tops, glaciers, Arctic icebergs through the tale, this is somehow even odder.) None of this strongly supports Greer's case. It seems to me equally plain that all of this scene-setting is, precisely, a manifestation of the varieties of consciousness in play. So right from the outset of sf, if we allow Aldiss's originary conceit, we see the hidden search for consciousness—not just of an intense cerebral kind, but manifested in the novel's several minds, not least the visceral and emotional agonies of both Frankenstein and his creation.

* * *

What Is It Like to Be a Double?

1886 Robert Louis Stevenson, *Strange Case of Dr. Jekyll and Mr. Hyde*

Scottish travel writer and novelist Robert Louis Stevenson was born in 1850 in Edinburg and died 46 years later in Samoa. His middle name, often erroneously pronounced "Louie," was given to the infant as "Lewis." This uncertainty in nomenclature is curiously reflected in a doubled identity explored in the proto-sf novella about the drug-induced consciousness and physical alternation between upright, priggish Dr. Henry Jekyll and his alter ego, wicked Edward Hyde. Jekyll's cruel and evasive double enacts, with conscience-less cruelty, urges that have always been hidden (hence, of course, "Hyde"). We never learn what disgraceful impulses Jekyll has indulged and kept secret for the sake of his reputation, but some critics find verbal evidence in the story that this was sexual engagement with other men—in Britain at the time, a criminal "perversion" punishable by death.

Stevenson's London is choked by the detritus of smog from burning coal, fouling the skies with metaphor. Jekyll's rather dull bachelor lawyer, Mr. Utterson, the story's main viewpoint character, beholds

> a marvelous number of degrees and hues of twilight; for here it would be dark like the back-end of evening; and there would be a glow of a rich, lurid brown, like the light of some strange conflagration and here for a moment, the fog would be quite broken up, and a haggard shaft of daylight would glance in between the swirling wreaths. The dismal quarter of Soho seen under these changing glimpses, with its muddy ways, and slatternly passengers, and its lamps, which had never been extinguished or had been kindled afresh to combat this mournful reinvasion of darkness, seemed, in the lawyer's eyes like a district of some city in a nightmare…

Iterations of imagery can be pressed into support for the guilt-racked repression theory, as they are in Elaine Showalter's *Sexual Anarchy* (1990). Hyde's domain, she claims, is a "representation of the homosexual body" with "a series of images suggestive of 'anality'…" This is possible, but more in the eye of the beholder than in the text. Jekyll admits only that "Many a man would have even blazoned such irregularities as I was guilty of: but from the high views that I had set before me I regarded and hid them with an almost morbid sense of shame." He acknowledges only that "my pleasures were (to say the least) undignified." Clearly, these "irregularities," boasted by many others, cannot bear the weight of shame projected on the love that dared not speak its name.

What we do see repeatedly is an instinctive horror everyone feels in the presence of Hyde, who is dwarfish where Jekyll is tall, and apparently younger than his mature counterpart. As we noticed in Mary Shelley's account of Victor's Monster, physical unattractiveness is somehow a mark of worthlessness and indeed moral malignity. We are told that Hyde

> "is not easy to describe. There is something wrong with his appearance; something displeasing, something down-right detestable. I never saw a man I so disliked, and yet I scarce know why. He must be deformed somewhere; he gives a strong feeling of deformity, although I couldn't specify the point."

Even Jekyll suffers this reaction, as self-loathing:

> This familiar that I called out of my own soul, and sent forth alone to do his good pleasure, was a being inherently malign and villainous; his every act and thought centered on self; drinking pleasure with bestial avidity from any degree of torture to another.

This mark of Satan is given a scientific explanation by Stevenson. Henry Jekyll has conducted chemical trials in his cabinet; "a particular salt which I knew, from my experiments, to be the last ingredient required" had been purchased "from a firm of wholesale chemists." Later, to his dismay, he learns that this key ingredient must have been contaminated by an unknown "impurity."

While his supply remains, he finds that he is growing addicted to the drug, in a way more distressing than mere tolerance. He is "cursed with my duality of purpose," opened to inspection and exercise by his concoctions. "I had now two characters as well as two appearances, one was wholly evil, and the other was still the old Henry Jekyll, that incongruous compound of whose reformation and improvement I had already learned to despair." Severed and revealed separately by his scientific intervention, his doubled consciousness is not just a proto-Freudian speculation but a real sundering of the self.

Jekyll is reduced to a kind of laboratory demonstration of the human mind's divisions of consciousness:

> It was a thing of vital instinct. He had now seen the full deformity of that creature that shared with him some of the phenomena of consciousness, and was co-heir with him to death: and beyond these links of community, which in themselves made the most poignant part of his distress, he thought of Hyde, for all his energy of life, as of something not only hellish but inorganic. This was the shocking thing; that the slime of the pit seemed to utter cries and voices; that the amorphous dust gesticulated and sinned; that what was dead, and had no shape, should usurp the offices of life. And this again, that that insurgent horror was knit to him closer than a wife, closer than an eye; lay caged in his flesh, where he heard it mutter and felt it struggle to be born; and at every hour of weakness, and in the confidence of slumber, prevailed against him, and deposed him out of life.

This is a realization more frightful than the growing awareness by Victor Frankenstein that his monster is, after all, just a transformed manifestation of himself, his self-crucifying consciousness. Decades later, these divisions would be explored in the first gaudy adventures by Edgar Rice Burroughs—not only the Tarzan saga, where a man's consciousness is modified and enhanced by his upbringing from infancy at the hands of apes (or perhaps an enduring troop of *Homo erectus*), but now a whole world of imaginary Martians. Meanwhile, H. G. Wells proposed another grisly scenario, this one another abuse of scientific discovery: the *uplifting* (as it would come to be known in the sf megatext) of "lower" species to the consciousness status of humans, if only barely—and with predictable gruesome consequences.

* * *

What Is It Like to Be a Beast-Man?

1896 H.G. Wells, *The Island of Dr. Moreau*

We have accepted Mary Shelley as in some respects the founder of science fiction in the post-religious mode, and Jules Verne as an adventure writer whose sf-like fiction borrowed bold ideas current or pending in his time (balloon flight and primitive submarines already existed, for example). Even so, it is plausible that the early fiction of Herbert George Wells (1866–1946) went systematically where nobody had gone before: a time machine capable to going forward and backward in time as it is experienced by the rest of us (1895), human invisibility (1897), invasion of our world by Martians delivered in spacecraft and razing the habitations of humanity in immense war machines (1898), tanks and atomic bombs (1903, 1914), and in some respects most grimly the transformation, by the rogue biologist Dr. Moreau (1896), of large animals into parodies of conscious humans.

Wells had training in Darwinian biology, but his time was well prior to the discovery of the DNA code so his vivisection nightmare could not use microsurgical genomics to modify existing species or create new ones. It might seem odd that a socialist like young Wells with an immense appetite for the new wonders of experimental and observational science should present what amounts to a horror story about its abuse. But of course as the Victorian era drew to a close, the political consciousness of working class people like Wells was afire with the perception that their kin were largely seen as fodder by those comparatively few aristocrats and upper class bureaucrats in charge of the Empire, with nothing to offer those *real* humans but life-long drudgery.

The Time Machine, with its effete and edible Eloi and brutish predatory Morlocks was an early parable of this division. *The Island of Dr. Moreau* carried it further toward the imperial and totalitarian future Wells sensed some decades into his own future. George Orwell's *Animal Farm* and *Nineteen Eight-Four* would carry these intimations still further, in harsh, wide-eyed fable and equally wide-eyed realist dystopia. In this sense, all these works of *fantastika* engaged in consciousness-raising, but not necessarily studies of consciousness as we discuss it in this book. *Moreau* certainly considers both.

Edward Prendick, whose education seems like Wells's, is lost at sea in 1887, saved and revived by an alcoholic medico, Montgomery, reviled by the foul-mouthed drunken captain whose ship carries animals of every kind plus a "black" man who seems as much beast as savage. As they approach their secret destination, the captain casts Prendick loose without water or food after Montgomery refuses to accept him on the island. The medico changes his mind, saves Prendick once more, and this grudging act of humane kindness prepares us for the contrary of such consideration for the animals who are fetched here for experiment by Dr. Moreau.

One of these animals is a cruelly caged female puma, whose unanesthetized and ongoing mutilation is so distressing that Prendick flees the crude research center for the surrounding forest. There, blundering as night falls, he finds other half-human half-beast creatures who eventually, but not until well into the novel, prove to be attempts at what later science fiction dubs "uplifting": in this case surgical modification to emulate the human phenotype.

Before he understands this, Prendick assumes that Moreau has *reduced* human victims to a quasi-animal state. "They were men—men like yourselves, whom you have infected with some bestial taint, men whom you have enslaved, and whom you still fear" (p. 90).

Even when informed of the truth—that these are enhanced animals—Prendick and we readers must ask: why are Moreau's unfortunate patients denied anesthesia (admittedly, in 1887 still not widely used even in surgery)? Does he derive sadistic pleasure from the infliction of pain? He denies this, claiming that mastery of pain is required by the true scientist.

> "I went on with this research just the way it led me... I asked a question, devised some method of getting an answer, and got—a fresh question... The thing before you is no longer an animal, a fellow creature, but a problem... To this day I have never troubled about the ethics of the matter. The study of Nature makes a man at last as remorseless as Nature" (p. 102).

Perhaps this is intended by Wells as a rebuke to the scientistic cast of nineteenth century medical research, for which the subjection of nonhuman animals is seen as irrelevant to ethics precisely because, as Descartes had taught, animals were held to lack true consciousness. And yet in many cases Moreau had carved and stitched fake humans until they possessed the power of speech and devotion. Prendick is astounded and dismayed to enter a cave where the rejected hybrids are catechized by the Gray creature in the Law:

> "Not to go on all Fours; *that* is the Law. Are we not Men?"
> "Not to suck up Drink, *that* is the Law. Are we not Men?..."
> "*His* is the House of Pain..."
> "*His* is the Hand that Wounds."
> "*His* is the Hand that Heals."

And then, a grandiose Moreau-based cosmogony:

> "*His* is the lightning-flash... *His* is the deep salt sea... *His* are the stars in the sky." (pp. 80–81)

Prendick's harrowing year on the island yields an ambivalent lesson: he is revolted by the Gulag cruelty and denial of agency or interest, yet he sees the mutilated creatures as less significant than genetic humans, even if they speak a barbarous version of English and discover this variety of authority structure. Perhaps the Beast-Men's failure to engage fully with their abominable situation is due to their tendency to revert to their species behaviors and instincts.

Moreau has blended diverse species. One becomes Prendick's companion: a St. Bernard dog transformed to a human shape. Others include a flesh-eating Hyena-Swine whom Prendick is forced to shoot dead, the gray Sayer of the Law who chants Moreau's Kipling-like teachings, the Sayer's supporter the Fox-Bear Witch, numerous others including a pre-Tarzan Ape-Man who sermonizes his "Big Thinks" and quite clearly possesses the qualities required to meet the criterion of "What it is like to be—," even if some of the lesser constructs seem more truly bestial.

The novel introduces Prendick to all these luckless Yahoo-like composites, until finally he is saved following Moreau's and Montgomery's deaths (the vivisectionist slain by his puma-woman victim) and the loss of the research center in a fire he himself causes by accident. Naturally his rescuers believe nothing of his fantastical experiences, and he withdraws into silence to avoid the fate of the madhouse. Like Gulliver in his final return from the land of the sublime equine Houyhnhnms, although for the contrary reason (he expects humans to revert to beasts), he can longer tolerate human companionship, and withdraws to the country where he writes his record.

* * *

What Is It Like to Be a Martian Brain-Transplant?

1927 Edgar Rice Burroughs, *Master Mind of Mars*

Consider the episodes near the opening of Burroughs' proto-sf, surely inspired by *Frankenstein* more than a century earlier. Ulysses Paxton, an American soldier near death in the First World War, hurls himself to the red planet Barsoom (as the Martians call their home world) via a kind of extremely solid astral projection, whereupon he arrives with his explosion-amputated legs repaired. This is the frankly *fantastic* part; the *science*, of an early twentieth century kind, describes not only the planet's several races of egg-laying humans, and formidable creatures with supernumerary limbs and sharp minds, but also a red-skinned Martian neuroscientist who is proverbially cold and single-minded with the know-how to swap brains between corpses and then reanimate them.

Ulysses, stranger in a very strange land, immediately encounters this old, naked and emaciated Martian male who takes the human under his wing (not literally, since the Barsoomian hominins are not avian despite their oviparous origins). This is Ras Thavas, in earthly terms some 2000 years old and the planet's titular genius, a magisterial experimenter. His immense laboratories contain a host of carcasses on shelves, and Ulysses watches aghast as the surgeon removes the blood from a hideous old woman, replaces it with a clear liquid, and carefully removes the brain. A radiantly beautiful young corpse is likewise trepanned, drained of preservative, replenished with the warmed blood of the crone, and the brains are exchanged. Soon the wealthy and rejuvenated termagant, Xaxa, returns to her high estate, but Ulysses finds himself captivated by her tragic surgical donor.

Now Burroughs dramatizes the problematic of mind, body and brain. Ulysses has never heard the woman speak, but he intuits from the physical beauty of her former habitation that she is surely sweet, kind, good-natured. Her new voice is shrill and harsh, but he tries "to forget those strident notes and think only of the pulchritude of the envelope that had once graced the soul within this old withered carcass" (p. 28). Does this imply that physiological beauty creates a generous personality, or does the causation work in the opposite direction? There is, in this case, no implication of spiritualistic duality. The transplanted mind is plainly an effect of the living brain. This assumption gains support when the sinister surgeon transplants half a human brain into the skull of a giant Martian brute, and the corresponding half of the creature's brain is conjoined with the human's.

What happens? Well, the partial brains tussle for control, but finally it is the translated tissue that proves dominant. The luckless fellow now inside the beast's body is understandably horrified. Ulysses promises to do his best, when he has a chance, to reverse the indignity. Interestingly, the once beautiful woman, with whom Ulysses is now besotted, makes the best of her lot and declines to be moved to another lovely corpse. "This old body cannot change me, or make me different from what I have always been. The good in me remains and whatever of sweetness and kindness, and I can be happy to be alive and perhaps to do some good" (p. 29).

This sentimentalized romance is now, in the twenty-first century, very close to being more than a *gedanken*. After decades of successful organ transplants, we seem on the verge of medical experimenters moving the brain of a damaged and dying body into the evacuated skull of a healthy corpse on life support. One imagines someone like the late Stephen Hawking relocated into the body of a young man (or perhaps woman) rendered brain-dead by an accident, or perhaps a Kim Jong-Un or Donald Trump resurrected in the young, strongly-muscled body of an executed criminal.

* * *

What Is It Like to Be a Mutant?

1935/1939 Stanley G. Weinbaum, *The New Adam*

An early misfortune for the emerging field of science fiction was the death (by throat cancer, just a year and a half after his first story was published) of 33 year old Weinbaum (1902–1935), Half of his promising work was not published until after his death—in the case of this superhuman story, some four years later. It is precisely a fictional portrayal of varieties of consciousness, in the curious tradition best exemplified by J.D. Beresford (*The Wonder*, 1911, rev, 1917) and Olaf Stapledon (*Odd John*, 1935, the same year Weinbaum died). The *Encyclopedia of Science Fiction* aptly notes that it is "the kind of story in which the superman cannot adjust to normal humans, and suffers fatal solitude."

Edmond Hall is a genetic anomaly, with an extra joint in every finger and thumb and the capacity to attend to separate trains of thought and observation with each cortical lobe. His mother dies soon after his birth, and his lawyer father shows no fondness for his odd son, although he does him no active harm. Hall proves to be ferociously brilliant, inventing in weeks a kind of atomic power based on cosmic rays before the secret Manhattan Program was even begun. Telling business men that this discovery had better remain hidden or it would be used in the next war, he says bluntly: "Would you place hand grenades in the paws of all the apes in the zoo...? Neither shall I" (p. 46).

The novel traces his brief life's course through three epochs, from infancy and childhood in three chapters, through Book I: The Pursuit of Knowledge, in ten chapters, to Book II: Power, in twenty four. This numeric increase reflects Edmond's growing complexity and paralyzing sense of futility.

Hall's disdain for other humans—if he *is* still human, and not one of a new species—has been noticeable since his childhood, but by early adulthood he does not bother to hide his mocking superiority. He reflects that an individual man "may be intelligent enough, but the mob-man never" (p. 61). A burst of premature postmodernism afflicts him: "Things devolve on the point of view; this is the only absolute in the universe, being the ultimate denial of absolutes" (p. 99).

In maturity he is torn between the beautiful flapper Evanne (yes, this Adam has his Eve, whom he knew from his early schooling when she preferred his rival Paul). Later he discovers a mutant female equal, artist Sarah. He seeks happiness with the former (wooing her with a sort of mental compulsion), then the latter who bears his mutant child. Finally he returns to Vanny before dying young, from expired vitality, in existential acceptance of the meaninglessness of it all yet hoping that billion year Nietzchean circular time will permit improvements next turn around).

The puzzle of consciousness is not quite made explicit in this novel but it is everywhere exemplified. Hall is a kind of advanced version of Julian Jaynes' ancient humans prior to a supposed *Origin of Consciousness in the Breakdown of the Bicameral Mind* (1976); each brain hemisphere contains a separate center of awareness and thought. This bicamerality has a quite different valence from that proposed by Jaynes, however, because while the two selves of Edmond Hall routinely hold highbrow colloquy, and swap bad poetry with each other, they can merge in stressful situations and focus as one with laser intensity (a benefit not available to Stevenson's Dr. Jekyll and Mr. Hyde, who have to take turns).

Vanny actually senses this inner bilocation, detecting a spectral and rather terrifying doubling of her husband. A few others are also partially aware of this troubling and creepy twinning. We get the impression that a penumbral detection of his difference is a major reason why people in general avoid his company, regarding him with innate suspicion. Rightly so, perhaps; it is increasingly clear that he (and Sarah, and some others who show up eventually, as mutants do in *Odd John*), really are a new species, unable to breed with *Homo sapiens* and comprising a kind of apex predator on their ancestor species. His first encounter with Sarah strikes him "like an awakening crash to a sleeper, his twin minds fused" (p. 171).

> And even while his consciousness reeled to adjust itself to this astonishing presence, some impish brain cells in the background were grinning. "Dog scent dog!" he thought sardonically, and raised theoretical hackles. (p. 171)

Moments before, Hall had imagined a cyborg future: "A truly civilized man would be in effect a free mind in a body of machinery." And yet, crucially, "This free brain of ours lacking the instincts that are a part of body could see nothing of beauty, and to that extent is not a truly civilized being" (p. 170)

Does the New Adam feel any moral responsibility or even emotional attachment to the elder species? Repeatedly, he thinks of Vanny's "little mind" in a rather fond way, as he did with a cheerful monkey he acquired when young and named Homo. After all, what really is the difference between Homo the monkey and *H. sap.* the species? This recognition naturally fills him with a kind of disgust when he craves the physical grace and sweet coital embraces of his human wife, for is this not a sort of bestiality? Certainly Sarah shares this assessment, but then she is a cold fish.

What we see in this novel, then, is an operatic showdown between Reason (chilly and unengaged, allegedly) and Emotional Passion (burning with a feverish heat, risking despair if the desired Other fails to reciprocate). Sarah reproaches him:

> "You blanket your life with verbiage, and tuck it in soft and warm while about you the lightnings flash. You argue with your own reason and temporize with your body, and are in all ways unworthy of your heritage" (p. 246).

Edmond's rational self sees the justice, but no true meaning, in her words:

> [U]nderstanding was possible between them, but sympathy never... he reflected that between himself and Vanny, exactly opposite conditions obtained.... "It is true, then, that bodily things are far more than intellectual; the important elements are not the highest. The mental is not the fundamental." (p. 246)

Vanny's response to his estimate is heartbreaking: "See, Edmond, I traded my soul for the chance to understand you, only the price I had to offer was not great enough" (p. 247).

The narrative craft of this novel surpasses that of almost all the pulp sf available in the USA at that time, and serialization in *Amazing Stories* was delayed until 1943.[2] This was an epoch when mostly male readers complained when some sf stories had elements of romance; that soppy stuff was for girls, not hardheaded engineers and would-be atom scientists—and certainly not if it argued in favor of emotions dominating over reason. Granted, *The New Adam* had no lack of fancy superpowered atom inventions, in which niton (radioactive radon gas) did something highly scientific to the element lead, creating abundant usable energy as well as the risk of global nuclear destruction. It is hard to imagine, though, that this was sufficient to offset the novel's emphasis on Edmond Hall's dubious love-life and Mimì-like *La Bohème* death. Young Isaac Asimov, who emerged a few years later, did not set foot on this territory of poignant emotionality for many years, as in his sentimental tales "The Ugly Little Boy" (1958) and "The Bicentennial Man" (1976). Let's turn

[2] Some delving in the records by Ian Covell (whom I thank for sharing this background information) suggests that the book was well started before the end of the 1920s. He cites:

https://library.temple.edu/scrc/stanley-g-weinbaum-papers

Series 1: Manuscripts, undated

1/4 "The New Adam." Original drafts: two college notebooks (French, and English & Chemistry) with his collegiate notes. Apparently used during his sales travelling in 1926 for a Chicago chemical firm in Wisconsin and Minnesota as note books and contain on 64 pages his first draft of "New Adam." First sentence: "Anna Hall died as stolidly as she had lived." 12mo. Ink on paper undated

1/5 "New Adam." Holograph in pencil, decorative headings on four leaves. 23 leaves (24p.). Includes: Contents leaf (2p.), Proem(8p.), and of Book I, chapter 1 (12p.) and two pages of chapter undated.

Mr. Covell summarizes: A 64-page "first draft" written in notebooks from the late 1920s...

now to that very famous sf writer and see if we can discern his own long investigation of the consciousness of robots and other beings.

* * *

What Is It Like to Be a Bot?

1939–1986 Isaac Asimov, the Positronic Robots Sequence

The Russian-American prodigy and polymath Dr. Isaac Asimov (1920–1992) was especially proud of one historic accomplishment: applying rational engineering principles, he defanged the meme of feral, hateful, destructive robots. (Well, defanged it for a time. Then came *The Terminator* and many other murderous artificial intelligence memes even more toxic than the pre-Asimovian versions. One of them, sanctioned by Asimov's estate, was the deplorable mash-up movie using the title of his first short story collection on the topic: *I, Robot.* Even though some of the characters' names were imported into a very different setting dreamed up and co-written by Jeff Vintar, the narrative centered on a lethal uprising of bots that was simply impossible under Asimov's cognitive rules, dating back to the 1940s: the fabled Three (later, Four) Laws of Robotics.

Every science fiction reader knows, as does almost everyone in the world other than Jeff Vintar, director Alex Proyas and the film's four producers, that the positronic brains of Asimovian robots are imbued with a set of unbreakable rules and prohibitions. These, we understand, are formulated in intensely complex and difficult mathematics, but for ordinary purposes are summarized as Three Laws: First, never hurt a human or allow harm though inaction to prevent it, Second, obey human orders unless that conflicts with the first Law, and Third, protect its own existence unless that conflicted with either of the earlier laws. This seems air-tight for a moment, and Asimov got a lot of mileage out of stories and novels that show breaches of the Three Laws—except that he then ingeniously revealed other explanations for the apparent failures.

Now that we live in a world of actual automata and useful robots, with prospects of human intelligence-grade systems perhaps on the horizon, the inadequacy of these Laws has been argued convincingly by engineers and philosophers.[3] Asimov knew that programmed competence without comprehension would inevitable run afoul of human semantics unless his robots, or

[3] See, for example, https://singularityhub.com/2011/05/10/the-myth-of-the-three-laws-of-robotics-why-we-cant-control-intelligence

some of them, were augmented to the level of human and posthuman intelligence, capable of parsing ambiguous statements and orders.

An early story, for example, has a mobile robot disappear and evade capture because an irritated human told it to "Get lost." The sf satirist John Sladek made endless fun of these robotic quirks (in this brilliant two-word quip, for example: "Yes, 'Master',"), with the sociopath Tik-Tok, the poignant and exploited robot Roderick who is clearly smarter and more caring than his human abusers, and plentiful examples of Three Law fails. The essay by Aaron Saenz cited above, from a think tank that hopes to create Friendly AIs and robots even smarter than we are, notes: "We don't want robots to obey anyone, we want them to obey just the people who own them. Would you buy an automated security camera that would turn itself off whenever someone asked it to?" The creepy and indeed horrific implication here is that robots of the Asimovian kind would indeed be owned—slaves, however much "it is like to be" them, as represented in Brian W. Aldiss's "Super Toys Last All Summer Long," the basis of Stanley Kubrick's final movie *A.I.: Artificial Intelligence*, completed by Steven Spielberg.

Coinciding with Asimov's invention of the constraining Laws of Robotics, several other sf masters looked with a certain fear and trembling at other possible outcomes of robot regulation. Ray Bradbury, in "Marionettes, Inc." (1949), devised a now-painfully obvious *Twilight Zone*-style parable in which a man buys one of those new perfect emulations of himself so he can sometimes escape his oppressive wife. The new version decines to remain in the cold basement and replaces the original in the affections of his spouse—who, meanwhile, has also copied herself. Jack Williamson's important story "With Folded Hands" (1947)—later the basis for the novel *The Humanoids*, and chosen for inclusion in *The Science Fiction Hall of Fame*—introduces a well-intentioned plague of small, polite, faceless black worker/police who swiftly make the world safe for humans by removing everything harmful and improving the disruptive with swift brain surgery. These Mechanicals, or Humanoids, are linked in an interstellar hive mind of monstrous "benign" oppressiveness.

The canonical Asimovian instance of "Mechanical Men" development[4] is R. (for "robot," of course) Daneel Olivaw, a humanoid lookalike of his interstellar designer. Daneel has impressive powers of speech and deduction, and superhuman acuity and physical power, but remains reined in by the Three Laws. Eventually he thinks his way to a Zeroth law: "A robot may not harm humanity, or, by inaction, allow humanity to come to harm." But this

[4] This term preceded Asimov's usage, but his early robot stories told of the puzzles and solutions encountered by the pioneering staff of "U.S. Robots and Mechanical Men, Inc."

requires theories of humanity and harm that not everyone agrees on, here in the real world. The question becomes urgent, but I do not think Asimov ever fully answered it: must a robot, even so wonderful as Daneel, be no more than a philosophical zombie without consciousness?

The last two paragraphs of Asimov's hefty 1960 volume titled (with unconscious sexism) *The Intelligent Man's Guide to Science* asked "What achievement could be grander than the creation of an object that surpasses its creator?" (*Opus*, p. 72). In the even more massive revision, *Asimov's New Guide to Science* (1984), he amplified this hope:

> With intelligences of different species and genera, there is the possibility at least of a symbiotic relationship, in which all will cooperate to learn how best to understand the laws of nature and how most benignly we might cooperate with them.... Viewed in this fashion, the robot/computer will not replace us but will serve us as our friend and ally in the march toward the glorious future—if we do not destroy ourselves before the march can begin. (pp. 866–87)

This utopian wish for a companionship with robotic intelligence does not specify AI consciousness, but that seems implicit. The ambiguity remains, and is never entirely clarified in the series of post-Foundation novels Asimov wrote in his last years, coordinating his several future histories (robots, galactic empire, psychohistory) into a single grand narrative. We learn that R. Daneel is not only intelligent and devoted to humankind (as he must be, by the Zeroth law now at work readjusting his telepathic platinum–iridium brain) but imbued with a consciousness rich with memories of 20,000 years shaping human and robot history. It is not *our* kind of consciousness, not quite; one finds an absence of greedy ambition, viciousness and self-importance, but also an unusual degree of candor in the service of well-considered goals.

At the end of the great sequence, Daneel reveals himself to a small group of humans, and one other:

> He was tall, and his expression was grave. His hair was bronze in color, and cut short. His cheekbones were broad, his eyes were bright... Although he seemed sturdy and vigorous there was... an air of weariness about him...
>
> His voice was utilitarian rather than musical, and there was a faint air of archaism about it...
>
> "I greet you all in friendship," he said—and he seemed unmistakably friendly, even though his face continued to remain fixed in its expression of gravity... "I am a robot. My name is Daneel Olivaw." (*Foundation and Earth*, pp. 490–91)

Trevize, the human explorer chosen to decide the fate of humankind, agrees to support Galaxia, essentially a hive-mind extending throughout the local Milky Way galaxy. The consciousness of Gaia, seed of that promised coalescence, finds Daneel's mind "neither quite human nor quite robotic." Meanwhile, a child awaits with them, yet another variant of self, with "brooding eyes... hermaphroditic, transductive, different" and unfathomable (*Foundation and Earth*, p. 510).

So at Asimov's signing-off, the shaping of life and mind in the cosmos is not, after all, secretly robotic, not governed by "what it is to be a bot," but rather the emergence of a galactic consciousness, growing to encompass the whole planet Gaia and eventually, it is expected, not just the Second Empire predicted by Hari Seldon, not just the Galaxia of the Milky Way, but the entire universe. Perhaps that is a consummation devoutly to be wished—but it might very well be read by humans in the twenty-first century as a program for the extinction of what is central to our worth as a species. As usual with this topic, that is a discovery to be made by each reader, and reconsidered every now and then as the evidence for the nature of consciousness becomes clarified.

* * *

What Is It Like to Be an Encrypted Supermind?

1939–1950 A.E. van Vogt, *Space Beagle* Stories; 1942–77 "Asylum" and Extended as *Supermind*; 1948–56 *The World of Null-A, The Pawns of Null-A*

In a 2018 Facebook essay tracing his own history in science fiction, Hugo-winner Alexei Panshin commented admiringly about

> the early stories of A.E. van Vogt—the once well-known but presently under-appreciated science fiction writer whose prose was lumpy and full of holes but whose concepts and images underlie *Star Trek*, and *Alien*, and *Dr. Who*, as well as any number of Philip K. Dick movies, plus Marvel Comics super-heroes, not to mention Japanese anime, manga and video games—documenting his fictional exploration of the idea of an emergent higher order of man.[5]

[5] Panshin, *Following My Nose*, May 21 2018.

Those stories by a Canadian writer first appeared in *Astounding* and other magazines as the Second World War was already grinding across Europe in 1939 but would not engulf the USA for another two years. Alfred Elton van Vogt (1912–2000), known generally as Van, was then already a 27 years old romance writer. His first sf appearance, "Black Destroyer," had a tremendous impact. Cover story in the July 1939 *Astounding*, it featured alongside Isaac Asimov's "Trends," that author's first sale to young editor John W. Campbell. The other contents were less memorable, but that issue is widely regarded as the launch of the fabled "Golden Age of Science Fiction."

"Black Destroyer" and its December successor "Discord in Scarlet," would be combined later with two other sequel stories as *The Voyage of the Space Beagle*. (I have always regarded that as an unfortunate title. For some time as an adolescent eager for sf, despite finding *Space Beagle* shelved with adult fiction I declined to open it, supposing that it must be an anodyne tale for small children. That proved very much not the case. A later paperback edition briefly changed the title to the banal *Mission: Interplanetary*, but generally it would retain its homage to Charles Darwin's epochal voyages of discovery in his British Majesty's survey barque The Beagle.)

Two aspects of consciousness recur as themata in many of van Vogt's works: manifestations of uncanny abilities of the sort Panshin rightly categorizes as a drive for transcendence (an argument developed at length in his and his wife Cory's history of the Golden Age, *The World Beyond the Hill*), and extravagant studies of alien and metahuman cultures of roughly the kind later appropriated by *Star Trek* in its search for new kinds of experience and unusual ways of life and thought. The enormous starship *Space Beagle*, and other gigantic vessels in later novels such as *Mission to the Stars* (a.k.a. *The Mixed Men*), are strikingly similar to interstellar craft in television and movies of the last 40 years. With total control of inertia they can perform a full-stop in the midst of superluminal flight. Some craft have internal teleportation (as does the *Enterprise*, though it is rarely used) to speed crew movements from one deck or end of the vast ship to another.

More strikingly, though this had also been foreshadowed by E.E. "Doc" Smith's sagas, alien cultures are graphed against social-historical models. Some of these were recurrent long cycles of the Spengler and Toynbee variety. The Japanese archaeologist Korita can identify the status of monsters they encounter: the final cyclical state of the fellahin, or peasant, with certain typical behaviors, limitations and strengths, inflexible caste systems, the "winter" period of advanced humanity with its Enlightenment demands for knowledge rather than superstition, and equality instead of hereditary rule, etc. It might be argued that such historical analyses depict on the coarsest scale the consciousness of each culture, with its dominants and its rebels.

At a different granularity, focused on the link between semiotic processes and individual awareness, van Vogt was enthralled by the General Semantics of Polish-American polymath Alfred Korzybski (1879–1950). Most of what he did with this body of half-baked thought is in its turn not only half-baked but used repeatedly as a plot coupon or get-out-of-narrative-trap cards. This was made entirely explicit (if rarely comprehensibly) in the diptych *The World of Null-A* and *The Pawns of Null-A* (1948–1956), where null-A stands for non-Aristotelian logic, allegedly a superior way to understand and manipulate spacetime and clarify the self. Here the emergent superman Gilbert Gosseyn (go-sane) not only has two brains, the better to think with but also to teleport objects under his inspection, plus the power to shift from one dying instantiation of his identity to a replacement clone waiting unconscious in a prepared hiding place. All of this huggermugger is exciting and provides a sense of mysteries on the very edge of one's mental grasp, especially if one is a clever but gullible adolescent at the time.

Similar but more imaginary corpora emerge in fix-up novels (built from shorter works, as *Space Beagle* was), such as the "logic of levels" in *The Silkie* (1969) in which modified humans can switch their bodies from human to aquatic to spacefaring, with suitable mental accommodations. Other marvels of thought are embattled in *Mission to the Stars* (1953), a fix-up of very early stories from 1943–1945 that features "Dellian robots" who are somehow hybrids of classic Earthly hominins and androids, which are not, of course, robots of the Asimovian kind. Depressingly, but perhaps understandably, van Vogt was drawn from these dreamy or nightmarish scenarios of ultimate power and success into his *Astounding* and *Unknown* colleague L. Ron Hubbard's bogus training systems. Eventually Van's own consciousness reawakened and he tired of that distraction, but really never recovered his youthful blazing intensity or crazy ingenuity.

The most vivid expression of van Vogt's yearning for a sort of secular spiritual expansion, although admittedly crudely wrought, is a climactic scene in *Supermind*, a fix-up based on his vampires-in-space story "Asylum" (*Astounding*, 1942). This is not so much an altered state of consciousness (although it is certainly that), but an altered state of encrypted being. An intelligent human discovers himself to be a mere shell identity for one of the Great Galactics whose lifestyle included the duty of keeping innocent worlds such as ours free from the voracious and highly intelligent Dreegh vampires. Here is the pivotal moment, which captures something about the human desire for full consciousness that cannot be rendered except in the lurid brilliancy of science fiction:

> —concentration. All intellect derives from the capacity to concentrate; and, progressively, the body itself shows *life*, reflects and focuses that gathering, vaulting power.
>
> One more step remains. You must see—
>
> Amazingly, then, he was staring into a mirror. . . and there was an image in the mirror, shapeless at first to his blurred vision.
>
> Deliberately—he felt the enormous deliberateness—the vision was cleared for him. He saw. And then he didn't.
>
> His brain wouldn't look. It twisted in a mad desperation, like a body buried alive, and briefly, horrendously conscious of its fate. Insanely, it fought away from the blazing thing in the mirror: So awful was the effort, so titanic the fear, that it began to gibber mentally, its consciousness to whirl dizzily, like a wheel spinning faster, faster.
>
> The wheel shattered into ten thousand aching fragments. Darkness came, blacker than Galactic night. And there was—
>
> Oneness! (pp. 62–63)

* * *

What Is It Like to Be an Enhanced Dog?

1944 Olaf Stapledon. *Sirius*: *A Fantasy of Love and Discord*

We have seen that philosopher Thomas Nagel, in his search for the nature of consciousness, would not be satisfied with learning what it *was like* to be *a human emulating a bat.* The hereditary bat genome constructs a very distinctive body and range of abilities and urges and typical behaviors. Humans can experience only a parody of this being-in-the-world, even if the exact neurological processes of a bat being a bat were encoded and ported into a human brain as a kind of hyper-dream or virtual reality.

What, though, if an animal phylogenetically closer to *Homo sapiens* (apes, say, or dogs) were shaped by selective breeding and pinpoint mutations to allow the creature to share a mutual if inadequate human *Weltbild*? If a lion could speak, Wittgenstein claimed, we would not understand him—but is that necessarily true? Suppose a developing puppy's brain could be tweaked in the direction of speech, and its tongue and jaw modified just enough to permit utterance—perhaps of a version of a familiar human language, even if it took especially patient attention from both speaker and listener to enable messages to pass back and forth between the species?

These days, advances in electronics and genetics might permit such modifications (if they were deemed unharmful), but in the global war years of the 1940s enough was guessed about radiation and its influences on chromosomes to allow a brilliant writer such as Olaf Stapledon (1886–1950) to take this admittedly implausible step in fiction. *Sirius* is a moving and ingenious science fiction novel published before atomic bombs were public knowledge, proposing that radiation might create the necessary physiological changes in a canine fetus to allow it to grow into a being able to live with humans as near-kin, despite its inherited infirmities—poor eyesight, lack of hands—and the need for extended longevity and a brain of human dimensions to provide human-grade raw intelligence (and powerful neck muscles to cope with the extra mass).

In the early 1920s, eminent Cambridge university physiologist Thomas Trelone combines these genomic elements plus hormones injected into the maternal bloodstream. By sheer persistence the scientist creates big-brained super-dogs, although most of each litter perishes. Finally Sirius survives, a blend of Border Collie and "Alsatian" (known these days as a German Shepherd), who was "a healthy and cheerful little creature that remained," like humans, "a helpless infant long after the other litter were active adolescents [. . .lying] on its stomach with its bulgy head on the ground, squeaking for sheer joy of life" (p. 17). By a stroke of luck, Trelone's wife Elizabeth bears a fourth baby at this time, a charming girl they name Plaxy. Sirius and Plaxy are raised together, like siblings, providing the superdog a perfect environment of love, fun and learning.

We find out about this highly unusual upbringing from a memoir by Plaxy's human lover and later spouse, Robert, who struggles to master his jealousy of the bond between enhanced animal and woman. His admiration for Sirius wins him over, and allows us to appreciate with almost total candor the powerful love and companionship of these quasi-siblings. As Britain plunges into a Second World War, Sirius has become an excellent custodian of sheep but yearns for a richer life. Bigoted and superstitious locals, led by a vindictive non-conformist minister, spread the rumor that Sirius is possessed by Satan. (Obviously his special capacities must be implanted directly by either God or Satan, since nobody imagines a human scientist could create such a hybrid, and it's much more satisfying to blame Satan and cry out for the animal's persecution and death.) "Salacious rumors" spread, accusations of "unnatural vice" (p. 164).

Are dog and human actually involved sexually? It seems likely, although Robert is indirect in dealing with the issue. Only "after many talks with Plaxy, that I realized how intimate her relations with the dog had become. The discovery was a shock to me, but I took pains not to betray my revulsion" (p. 169. However, despite his initial outrage, he "could not but realize that the passion which united these two dissimilar creatures was deep and generous." It is the spiritual nature of their union ("spiritual" not in any traditional religious sense) that explains their mutual devotion. Plaxy, "so she told me long afterwards [found] that in those strange, sweet moments she was taking the first step towards some very far-reaching alienation from her own kind. Yet while they lasted they seemed entirely innocent and indeed beautiful." Sirius himself notes: "Only in the most articulate, precise self and other-consciousness was the thing to be found."

It is this mannered but emotionally engaging struggle to know one another's consciousness and share it to the limits of possibility that Stapledon achieves his greatest success with this novel. Brian Aldiss called it, with his characteristic paradoxical twist, "the most human of all Stapledon's novels" (*Trillion Year Spree*, p. 198). It is true, though, given that the most notable works by this Marxist visionary are cosmogonic voyages into the depths of space and time and mind, one revelation upon another, until the cosmos is revealed as a kind of immense slow conversation between parts of a god. Even his marvelous and chilling novel of human mutation, *Odd John*, is less emotionally engaging than this wondrous love story of a human and an enhanced dog. It is inevitable, and tragic, that Sirius must die from gunshot wounds, victim of both blind stupidity on the human side and elements of his own not quite controlled bestial rage.

Here is a kind of answer, then, made long before the question was raised by Nagel, to the puzzle of sharing consciousness with a non-human. Yes, it might be possible—if the mind of the Other is first shaped and then nurtured to resemble that of a human, and then if the innate boundaries are acknowledged and respected. This is a theme to which we shall return in other science fiction stories and novels, perhaps most notably in John Crowley's *Beasts.*

* * *

What Is It Like to Be a Chess Piece?

1946 Lewis Padgett, *The Fairy Chessmen* aka *Chessboard Planet*

"The doorknob opened a blue eye and looked at him."

For many years, this hallucinatory start (and ending) to an *Astounding* magazine serial by husband and wife team Henry Kuttner and Catherine Moore, writing as Lewis Padgett, was applauded as perhaps the most startling in all science fiction. Some two decades later, it seemed a commonplace of psychedelic hijinks, so this just-postwar psychothriller was passé to the point of being forgotten by newer fans. A further half century on, it is almost unheard of. Yet the Kuttners, who would shortly abandon sf and other fiction for the formal study of psychology, were pioneers in prodding at the roots of consciousness.

Much earlier, the boy genius Arthur Rimbaud had endorsed derangement of the senses:

> A Poet makes himself a visionary through a long, boundless, and systematized disorganization of all the senses. All forms of love, of suffering, of madness; he searches himself, he exhausts within himself all poisons and preserves their quintessences.

While Civilian Director of Psychometrics Robert Cameron is no poet, his senses are flung into confusion, as are those of many research scientists and warriors in a future hot war between the West and the Eastern (almost certainly Japanese) Falangists, who suffer from a "racial psychological handicap, perhaps," whereby failure demands "honor suicide" (pp. 103–04).

When Cameron looks up from his desk at an ordinary office clock, it *speaks* the hour. When he clutches the eyeballed doorknob, it has become half-solidified jelly. An altimeter smiles at him. Later, he is convinced that his flesh is covered in bugs. A glass of water suddenly tastes foul. He finds an egg in his locked safe; then it is gone. The edge of a spoon thickens and kisses him on the mouth. All these psychotic anomalies appear to be related to an equation that cannot be resolved or countered, driving those attempting to circumvent it into madness or worse.

He and his confidential secretary Ben DuBrose visit a secret sanatorium to view troublesome Case M-204, and find the patient floating five feet above his hospital bed, convinced that he is Mohammed halfway between earth and heaven. Little wonder that the original title of this two-part serialized novel was

The Fairy Chessmen, a term from 1914 referring to chess played with unusual pieces or modified rules, such as dazzlingly difficult three-dimensional chess played on a special set of boards with additional pieces such as the grasshopper and nightrider, and aberrant geometries.

It turns out that various fundamental physical constants are now variables, at least as perceived by some. The minds of people trying to adjust to this loss of familiar empirical and cognitive moorings become unhinged. Consciousness has gone astray.

Cameron's aide and alter ego, Seth Pell, is four years older than DuBrose's 30, an astute, daring thinker who declares himself a "misogynist" (he clearly means "misanthrope," although it is notable that very few women appear in the cast other than Cameron's wife Nela). Perhaps the most shocking and bleakly realistic moment is a videophone conversation between Seth and a Dr. Emil Pastor, a physicist trying to unlock the mystery equation. Pastor has done so, to his great satisfaction. His discovery (which perhaps only holds for him) is that reality is hollow, an illusion, and he can shape it by intention. In proof, he causes Seth to vanish. If life is an illusion, this at least is not; Seth is dead.

The narrative largely follows DuBrose and his increasingly disabled Director. Cameron is plagued by hallucinations that seem focused on him by enemy operatives of the Falangists, who have mastered the equation. The general explanation for the insanity afflicting top engineers derives from their increasing and finally intolerable sense of failure in this dreadful exercise in cognitive dissonance. These intense and intent experts cannot cope with an equation in which fundamental forces can change from moment to moment. Following a hunch, DuBrose locates a genial and relaxed mathematician, Eli Wood, an accomplished player of Fairy Chess. Only a mind like his might combine mental agility and expertise in the discipline required to understand the equation. (Probably the latter is meant as a kind of extension of Schrödinger's Equation and other arcana of quantum physics, perhaps of the sort promulgated in later decades by the "wooly masters" of new age pseudo-thought.)

> "...your equation is founded on the variability of truths... If mutually contradictory truths exist, that proves they are not contradictory—unless," he admitted mildly, "they are, of course. That's possible too. It's simply fairy chess, applied to the macrocosm." (p. 79)

Complexities ensue, largely uncontaminated by literary characterization of the sort effortlessly driving Olaf Stapledon's engaging *Sirius*. The source of the equation is revealed by a mad young man, Billy Van Ness, to be a totalitarian warrior society of the very far future. A government courier, Daniel Ridgeley, proves to be an enigmatic visitor and supersoldier from that future. This thread of the novel anticipates the time travel paranoia inflecting much of the fiction then yet to be written by Philip K. Dick, who surely learned some of his craft from the Kuttners; he was 18 when the serial was published. Billy, like some other youths, lost his sanity at the age of 22, after being bathed in babyhood by gene-targeted radiations from mysterious Domes arrived from the future. He has been mutated by these invasive procedures into an ETP, one with extra-*temporal* perception. He detects "wildly oscillating" duration, past and future (p. 71).

Meanwhile, Pastor is torn between the lure of absolute power and his conviction that he is now God, although not the previous God who died for humankind. He perishes offstage at the hands of Ridgeley, who in turn is rendered comatose by use of a counterequation only he and the Falangists understand:

> Everything that had been or was to be, Ridgeley perceived in a shifting, monstrous kaleidoscope that became clearer as his perception sharpened. It was not merely sight. ETP is something else, a consciousness of the objective that goes beyond vision and sound and hearing (p. 116).

Consciousness is, in fact, the key to this strange little novel. Experience is deformed by the equation, when used as a weapon. The mind is driven into madness by denial of consistency and truth. Only a counterequation of the same generic type can defeat it, and the realization that humans will tear at each other forever until they have become the blond beasts that Nazism, only just defeated at publication in 1946, yearned after—unless our consciousness recognized that

> [t]he Enemy stood at the gates of the sky... The hostile universe that had always made man band together in a common unity... the stars... were the Enemy. The hostile, distant, alluring, secret stars. And they, too, would be conquered. But that would be no sterile victory.
>
> DuBrose thought: *The old order changeth, giving place to the new.* (pp. 122–23)

It is a perfect summary of the doctrine of John W. Campbell, editor and prophet of *Astounding*, proponent of the consciousness of the age of space and the new age of mental prowess once known as magic.

* * *

What Is It Like to Be Immature?

1948 Theodore Sturgeon "Maturity"

One of the early major novellas by Theodore Sturgeon (1918–1985) deals with the contributions to consciousness and creative imagination of unchecked endocrine hormones, and interrupted treatment. Written in early 1946, it first appeared in *Astounding* in February 1947, but was heavily revised for his collection *Without Sorcery* in 1948. The narrative is equally poised between Dr. Margaretta ("Peg") Wenzell and her patient, the dazzlingly inventive and buoyant Robin English. In an introduction to the collection, Sturgeon describes Robin as "one of the most captivating characters ever to take a fictional bit in his teeth".[6] At getting on for 29, he is just four days Peg's junior but seems in some respects like a clever child. His endocrinal disorder is hyperthymia, usually marked by charm, confidence, egocentricity, and rather driven sexual appetite (although this, while not absent, is not emphasized), and energy in abundance. He is, in short, the contrary of the tale's title: he is delightfully *immature.* This will change, leading to his willed death.

We observe Robin before treatment, surrounded at home by clutter, unfinished paintings, spiders and small animals, books everywhere, unwashed dishes piled in the sink, his mind endlessly observant, and then after treatment. The story is, indeed, exactly an exploration of **What It Is Like to Be a Hyperthymic Genius**. Specifically, as its author noted in 1979, it "was preceded by two years of research—research which consisted of asking everyone I met—young people, old ones, rich poor; strangers, loved ones, even faceless voices over the telephone: 'What is maturity'."

That is increasingly the quest of Robin English as well. At first he abides by treatments for his disorder, while suspecting that it will mute and dull his "immature" insights and creativity. When it becomes clear that Margaretta is in love with him, and resisting involvement (they never have sex), and that her colleague and mentor Dr. Mellett Warfield is in love with her and bitterly jealous of Robin on general principles, the young man abandons treatment and moves into a lifestyle of dazzling social and popular artistic success. It seems that his untreated condition is settling into a kind of wildly successful adulthood: his first novel, a fabulously successful musical with three songs drawn from it taking turns at the top of the Hit parade, a book of poetry, fame. Like a baroque figure from Alfred Bester's sf of the next decade, he wins a pie-eating

[6] Story Notes, *Thunder and Roses* (p. 340).

contest, records a collection of muezzin calls, develops a playboy persona to complement the shift in his appearance from small-chinned youth to handsome adult.

If this starts to sound like Sturgeon in the self-idealized grip of what is now called a "Mary Sue" or wish-fulfilment performance, that suspicion is rather confirmed by Robin's radio address "on the evolution of modern poetry which was called one of the most magnificent compositions in the history of the language" (p. 32). But really this is Sturgeon's whimsy; he was himself a notable drop-in spontaneous humorist on New York City late night radio with Jean Shepherd, and parts of his fiction were actually verse in disguised form. Here is an example from "Maturity":

> The year grew old, grew cold and died, and a new one rose from its frozen bones, to cling for months to its infantile fragility. It robbed itself of its childhood, sliding through a blustery summer, and found itself growing old too early. What ides, what cusp, what golden day is a year in its fullness, grown to its maturity? Where is the peak in a certain cycle, that point of farthest travel in a course which starts and ends in ice, or one which ends in dust, or starts and circles, ending in its nascent dream. (p. 50)

This is the map not of a calendar but of the flow of consciousness itself, Robin England's trail from boyish fascination with everything mechanical and *explainable*, and beyond that to *inventable*, ends in its nascent dream of maturity.

> "My brain isn't softening. It's—deepening. A Klein bottle has only one surface but can contain a liquid because it has a contiguity through a fourth direction; my mind has five surfaces, so how many different liquids can it contain at once?" (p. 56)

His final message to Peg and her future partner Mel is portentous and yet minimalism itself:

> . . .the simple wisdom he wrote; not a definition of maturity, but a delineation of the Grail in which it is contained:
>
> "*Enough is maturity*—" (p. 59)

Curiously, decades later a version of this slogan would be found in the title in eco-activist Bill McKibben's tract *Enough: Staying Human in an Engineered Age*. It is not at all clear that Sturgeon would have relished this (accidental?) appropriation, although some of his most vivid and memorable stories show us cultures where greed and the trappings of power and accumulation are despised—replaced, though, by unpretentious, beautiful and productive transhuman technologies. In any event, Robin instructs his grudging acolytes in his understanding of mature consciousness (p. 55):

"My mind is working on two levels," he said. "Maybe more.... It's too slowly to say it."

"Too hard to say it?"

"Too slowly. It isn't a thing you can say piece by piece. It's a whole picture; you see it all at once and it means something."

"I don't understand."

"No."

* * *

What Is It Like to Be an Alien's Dream?

1950 Theodore Sturgeon *The Dreaming Jewels*

In an admiring 1985 postmortem introduction to Sturgeon's final novel, *Godbody*, Robert Heinlein mentioned his friend's youthful circus ambitions—before succumbing, as so many did prior to antibiotics, to rheumatic fever. The disease "robbed him of his dearest ambition: to be a circus acrobat.

> In high school, by grueling daily practice, he had transformed himself from that fabled ninety-pound weakling into a heavily-muscled and highly skilled tumbler, one who could reasonably hope to join someday the 'Greatest Show on Earth'... He recovered... but with a badly damaged heart. A circus career was out of the question." (Heinlein, p. 8)

Despite his draft board 4F status, Sturgeon worked building airstrips as a civilian heavy-machine operator in the Caribbean. This grueling hot work could have killed him; it failed to, but it left him (as tuberculosis had left Heinlein) without much capacity for a physically demanding occupation. Instead, he made his living as a free-lance writer, often hobbled by paralyzing writer's block. These antinomies in his life's arc perhaps provide an explanation for his considerable capacity for probing the consciousness of thwarted men and women, often brilliant, who persist in the loneliness and darkness of disability or rejection—even though he himself was much loved and applauded.

This is the troubled ambiance of so much of his fiction, from "Saucer of Loneliness" and other short works through his tormented novels, especially *The Dreaming Jewels*, *More Than Human* (with its gestalt mind-meld), *To Marry Medusa* (in which humanity is technologically bonded into a hive-mind that overwhelms the galaxy), *Some of Your Blood* (where menstrual cunnilingus is the sacrament saving a desolated man and woman), *Venus Plus X* (a utopian enclave of surgically modified hermaphrodites).[7]

In the first of these novels, Horton, nearly nine years old, is a foundling, adopted for failed political purposes by the cruel and narcissistic Armand Bluett and his complaisant wife Tonta. Both torment Horty and resent his intrusion into their unpleasant life. On the first page of the novel, we are told, bafflingly, that "Horton's parents were upstairs, but the Bluetts did not know it" (p. 5). A live-in maid and her boyfriend, perhaps? No, only the three Bluetts are in the house. We learn fairly quickly, before Horty does, that his parents are a pair of glistening and conscious alien jewels attached as eyes to a revolting old jack-in-the-box, Junky, found with the abandoned baby. These beings possess the power to create, shape or reshape, and destroy organic creatures. With Horty they have made a biological error; he needs extra formic acid as he grows toward adolescence, and finds it in ants, which he licks up eagerly to the derision and revulsion of his schoolmates and teachers.

In a fit of rage at the lack of neighborhood respect likely to follow this discovery, Armand beats the child and flings him into a closet, slamming the door on his hand, severing three fingers. (We know from *Argyll: A Memoir* that this sort of treatment was familiar to Ted Sturgeon, treated badly by his stepfather.) Horty has had enough; he wraps up Junky and escapes from the house, chances upon his young crush Kay and bids her farewell, and dazed by blood loss and pain clambers ineptly into the back of a truck paused at traffic lights. By a strange coincidence, he is hauled on board by another creation of the Dreaming Jewels, a cigar-puffing midget named Havana, and meets still more: the kindly, deaf and repulsive Solum the Alligator-Skinned Man, and two miniature beauties, the albino Bunny and the especially gorgeous Zena who, oddly enough, somewhat resembles Horty. He has found the end of his loneliness and dispossession with these carnies, on their way to meet the carnival, under the command of misanthropic, former medico, hatred-driven Pierre Monetre, the Maneater.

This elaborate set-up, deployed in just twenty opening pages, steps us into a fairly typical Sturgeonesque comedy of consciousness and emotional need. No

[7] I have explored the second and third of these in *Psience Fiction* (2018), so will not repeat that here.

doubt Armand experiences himself as a worthy fellow unfairly beset on every side; his wife appears to have numbed her self-awareness just to get along. Horty (who halts his growth for years, to remain in the carnival) and Zena and other little people are profoundly *non-orthodox* in origin, which perhaps explains their almost supernatural caring for each other. Horty modifies himself physically in a kind of gender reassignment, becoming Kiddo, allegedly Zena's nineteen year old sister, while remaining masculine despite convincing disguise. And regrows his lost fingers.

We learn that Monetre, the Maneater, has found curious crystals that somehow cause duplicates of trees, herbs, and eventually animals and even humans—sometimes exact, although many of these manifestations are flawed or frail and die. In bursts of psychic malice he forces the jewels to do his bidding, but is unable to communicate with them. Perhaps if a perfect human creation might be produced, he or she might be twisted into the role of go-between. We assume that Horty is this chosen one. By now, he/she has worked for a decade in the Maneater's carnival as age-arrested Kiddo, secretly eidetic omnivorous reader but essentially feeble and directionless, under the wing of Zena who has now fallen in love with him/her. She sends him away for his own protection. Three years later, having grown some backbone and confidence, he tracks down the loathsome Armand. Now a judge, his adoptive father is paying unwanted sexual attentions to Kay, the young man's childhood love. Horty morphs himself into her likeness, agrees to a private meeting with Armand, and horrifies the lubricious and soon vomiting brute by chopping off three of his own fingers, again.

Tormented by the horror of a second amputation, Armand locates the carnival and presents himself to Monetre, revealing Horty's true nature but giving the mistaken impression that he is now Kay. The final third of this short novel proceeds more in the manner of the original "pulp fiction" conclusion of "Maturity," blended with exposition: Zena brutally battered but finding Horty (now tall, 100 lbs heavier and with manly shoulders), the two jewels that were Junky's eyes in her handbag, followed by hypnotized Bunny, and then Solum who steals the bag, and at last Zena is slain... But in the end we do share some of the insights and barely sayable experiences of Horton, who is the perfect dream of that pair of crystals. This is a genuine attempt at science fiction investigating consciousness—not just Horton's not-really yet more-than human mentality and modifiable body, but the inwardness of the jewels whose dream he is, who care no more for him and his purposes than we do for the evanescent dreams after we wake. Zena says,

> "Making those things is nothing to them. They live a life of their own. . .The things they make are absent minded things, like doodles on a piece of paper that you throw away" (p. 118)

For Horty, seeking their aid,

> "the blacking out of the ordinary world revealed another. It was not strange; it has co-existed with the other. . . These self-sufficient abstracts of ego were the crystals, following their tastes, living their utterly alien existences, thinking with logic and with scales of value impossible to a human being." (pp. 137, 142)

Delving like this into the impenetrable depths of an alien consciousness is perhaps no more than the commercialization of negative theology, where it is pointless, indeed laughable, to try to utter the unutterable. Yet we are humans whose hearts can break, so it is only appropriate that Zena returns from her cooling death at the scent of formic acid and that Horton puts on the physical form of the Maneater now killed by him, readying himself to retrieve and render harmless many fatal plague spores and poisons scattered across the routes of the carnival. What more can one demand of an entertainment with aspirations, in its way, to a metaphysics of unearthly consciousness?

* * *

What Is It Like to Be a Consciousness Seed?

1950 Katherine MacLean, "Incommunicado"

One of the most persistent and injurious forms of what might be called "consciousness bigotry"—often internalized in those subject to it via "false consciousness"—is the subordinated role of women. Notoriously, this cultural dichotomy reaches far beyond the obvious physiological differences. It extends, for relevant example, into typecasting and denigration of female scientists and authors whose skills and expertise often used to be scorned, unless they managed to cloak their gender, as "U.K." Le Guin was encouraged to do when *Playboy* bought a story from her in 1968, and Alice Norton wrote as the ambiguous Andre Norton.

A widely accepted story tells us that Katherine MacLean (born 1925 and still alive as I write) refused to take this route, even at the start of her sf writing career. Actually, according to sf scholar Eric Leif Davin, her family insisted mistakenly that no sf editor would accept a story with a woman's name on it, so she sent it in with her name shown as "K. MacLean." John W. Campbell at *Astounding* took it despite learning she was female, but required certain changes (as he often did with male writers too).

Her engineer-inventor father was supportive of her interests in science topics, and she held a laboratory job by the time she was 16. After the war, she received a BA in economics and MA in psychology, and finally became a hospital technician.[8] During this time she was writing, submitting and selling sf stories under her own name. After several novels were published in subsequent years, her novella "Missing Man" appeared in *Analog* and in 1972 won a prestigious Nebula award. But what of her most famous early story? It languished in limbo for three years after it was accepted, but not at first published, by Campbell.

According to the standard tale of chauvinism, "Incommunicado" was deemed obviously too savvy to be the work of a woman, so Campbell located her father and tried to bully him into confessing that he was the real author. Eventually persuaded of the truth, Campbell ran the story in June 1950, by which time MacLean had already placed another two stories with *Astounding.* Startlingly, Davin, in conversation with her, was told erroneously that Campbell insisted on changing the billing from Katherine to K. MacLean—even though in fact he had featured the story on the cover, with her name shown in full there and inside the magazine. This mythological rendering supported the legend that Campbell disdained women and either would not publish their work or required masculine pseudonyms.

So what was this troubling story about? Despite the somewhat feminine figure displayed on the cover, delicate fingers poised above an enormous keyboard, other hand held against the forehead, all the major characters are male, with one wife and two children glimpsed only in passing. A strand of tape loops across the page, musical notation written on one side, punched computer code on the other. Here is the Two Cultures of Lord Snow wedded together, significantly *avant la lettre.*

Drawing on both information theory, newly minted in the late 1940s, and musical/acoustics theory, the intriguing notion in "Incommunicado" was that the emergence of *Homo superior* might be spurred by instructing technicians and other smart people to reprogram those aspects of their brains devoted to parsing and generating meaningful sounds, Then they would stumble their

[8] See Brian Stableford, "Katherine MacLean," p. 487.

way to the use of these unheralded skills to expand consciousness via bleeps, rather than visual coded images, created by computers. Bearing in mind the absence of graphical user interfaces and displays in 1947 (not available until the early 1970s), this was not such an extravagant idea. Not that computers still make the kinds of eerie burps and buzzes once familiar from mid-century movies. Besides, it will always be true that existing human brain structures handle a very much larger data input visually than the ears can cope with. (Lester del Rey made this point in *Analog*, May 1975, to denounce the lovely conceit in M.A. Foster's *The Warriors of Dawn*, in which engineered humans called Ler can *sing* messages directly perceived as richly visual images.)

Building an orbital station near Pluto, a crew of techs uses specialized jargon, except for one frustrated genius who can intuit what steps to take just from the acoustics of the place and its computational machinery. But he cannot easily communicate with others without the help of his coordinator friend Mike. When Mike is killed in an accident with cascading consequences, inarticulate Cliff deserts his station because he no longer has any useful purpose.

His expertise is undoubted, but takes an unusual form not communicable to others. His hands are "strong, clumsy, almost apelike" but know "all the secrets of machinery by instinct... If all the lights of the sky were to go out, or if he were blind, he could still have cradled his ship to any spaceport in the system... it was instinct, needing no learning" (p. 120). Yet Cliff feels stupid, and with Mike's loss he believes he is incompetent.

At the station, he looks for someone to replace Mike, but young genius Archy, jazz musician aficionado and son of the late expert in computer indexing, turns him down coldly. Sounds and whistling and hummings pervade the station, embarrassing people as they too emit these unstoppable noises: Reep beep, *Reeb* beeb. Foo *doo*. At length Cliff finds the entire staff of the station cavorting about, blowing or plucking or beating instruments. He hears "a dreamlike clamor of voices surging up in his mind" (p. 139). The station's psychologist, en route back to earth for the birth of his child, uses mirroring to share Cliff's experience and explain that far from being stupid he is *an adjustable analogue*, in effect the next stage in evolution.

> "Your concepts aren't visual, they are kinesthetic. You don't handle the problems of dynamics and kinesthetics with arbitrary words and number related by some dead thinker, you use the raw direct experience that your muscles know." (p. 137)

Cliff understands at last that on Earth the citizens of a failing civilization struggle to make sense of its basic processes, "most of them baffled even by the simplicity of algebra, and increasingly hostile to all thought." Their consciousness, in short, is hamstrung, as Heinlein would argue in his novella "Gulf."[9] But unlike Heinlein, MacLean has a redemptive insight: the tedium or bafflement of their daily grind was made pleasant by music. People would spontaneously wince if, out of two hundred instruments, "a single violin struck four hundred forty vibrations per second where it should have reached four hundred forty-five." In a burst of twenty-first century prophecy, she notes that "even on the sidewalks people walked with their expressions and strides responding to the unheard music of the omnipresent earphones." Archy and Cliff understand that this orchestra of machinery and computation is altering and enlarging the consciousness of people who otherwise struggle to grasp the elements of formal learning.

It was no accident that consciousness was a key element in MacLean's fiction, and not just because of Campbell's and then MacLean's early intoxication with Dianetics. (Andrew Liptak claims that "she began to work with Campbell going through his system of logical exercises and lectures."[10]) As the late, major sf editor David Hartwell noted, in *Age of Wonders* (1984), writer and editor Judith Merril declared her conviction that, in sf at that time, she and Theodore Sturgeon and Katherine MacLean "were putting into print and into words ideas whose time was about to come, making it possible for people to become conscious of it... [We] felt that what we were producing was consciousness seeds, which were going to grow and expand" (p. 110). That, taken seriously, was the very essence of "Incommunicado."

[9] Heinlein, *Astounding*, Nov–Dec 1949.

[10] https://www.kirkusreviews.com/features/fantastic-foresight-katherine-maclean/

5

The Second Golden Age

The mind is only the processing focus of evolving organization, and this makes its role an instrumental one. In other words, the mind is not the "end" but only the "means" for the promotion of what the universe is all about... Thus, although the mind is not the hero of the epic, its role is crucial. It involves its working on the script, acting in the play, directing it, and being responsible for the production.

Zoltan Torey, *1999, p. 233*

What Is It Like to Be Your Own Twin?

1951 Wyman Guin, "Beyond Bedlam" *(Galaxy)*

We usually seem to ourselves to be integral, a unified self stretching back to our earliest memories. Psychology, neuroscience and simple observation reveals that this is a sort of delusion, as Daniel Dennett and Tor "User Illusion" Norretranders[1] proposed independently at the end of the 1990s. But can a mind, which is the ostensible victim of illusions, itself be an illusion? Not, perhaps, if the mind is indeed unitary—but it does not take much introspection or observation of friends and foes in different settings to notice the polyvalence of behavior and thus, arguably, of the partitioned self that gives rise to this multiplicity. The wanderings and lurches of a mad mind are just the

[1] https://archive.nytimes.com/www.nytimes.com/books/98/05/03/reviews/980503.03johnst.html

D. Broderick, *Consciousness and Science Fiction*, Science and Fiction,
https://doi.org/10.1007/978-3-030-00599-3_5

more extreme instances of the "divided self," as antipsychiatrist R.D. Laing dubbed it more than half a century ago.

In the last three months of 1950, editor H.L. Gold launched *Galaxy*, a marvelous and more sophisticated alternative to *Astounding*, gathering many of the best writers nurtured by earlier magazines and inviting good newcomers to compete for his competitive rates. One of the early finds was pharmacologist and ad executive Wyman Guin (1915–1989), whose *Brave New World*-ish story "Beyond Bedlam" was greeted with applause when it appeared in the August 1951 issue. Today that endorsement persists in comments by notable sf critic and encyclopedist John Clute: "a brilliant novelette" that explores a bifurcate future of universal mandatory schizophrenia "with literacy and verisimilitude" (*SFE*, 2018).

The key mistake in this assessment is that Clute, like Guin himself, thinks "Schizophrenia is where two or more personalities live in the same brain" [*Beyond Bedlam*, p. 150]). This is a careless and badly misleading statement that confuses "schizophrenia"—a chaotic and confused emotional and cognitive condition—with "multiple personality disorder" or more recently "dissociative identity disorder," in which alternative enduring selves manifest as distinct and different people. This latter is what Guin's future ensures with the forcibly imposed use of drugs, obliging all citizens to live a doubled life in sequential-day terms. This allegedly removes the brute animal tendencies of humankind toward murder and mayhem, and frees and clarifies the competing skills and talents of each Ur-individual.

Taken as a witty (if rather cruel) jape drawn out at some length, this *novum* has its merits. Guin takes the proposal for a stroll, as Robert Sheckley did around the same time with various silly notions and brisker delivery. What would such a life be like? Well, despite a rigid apartheid, suppose a man or woman in an unsatisfactory marriage (this was the early 1950s, after all) catches a glimpse of the spouse's alternative and falls passionately, uncontrollably and illegally in love. Wicked infidelity with one's own alternative spouse! Not quite knee-slapping mirth, but droll in the way that would become familiar to *Galaxy* aficionados.

Despite these deficiencies, the novelette does dramatize some of the issues we confront in our quest for understanding of consciousness. What would it be like to be not only you-prime (Guin's "hyperalter"), but also the subordinate "hypoalter" you? But that distinction is also a cause of contention in that world, where "there was occasional talk of abolishing the terms... because they were somewhat disparaging to the hypoalter, and really designated only the antisocial power of the hyperalter to force the shift" (p. 161). Only by choosing to "force the shift" could one self briefly enter and observe the other world with its entirely unknown inhabitants. This is plainly a future without microcomputer surveillance, but for our current purpose that small failure of prophecy is irrelevant.

The question is: can two distinct states of consciousness really be imagined as time-sharing a single body, and indeed maintaining parental and spousal roles without being tipped into genuine psychosis? It is the kind of question science fiction is especially apt to consider, along with "can a single consciousness inhabit five different bodies at the same time?" or "can a man and a woman switch their gendered bodies for a holiday?" or "can a time-loosened consciousness in a block universe focus its attention on different epochs of its self and *be there* in the past or future"? (This last seems to be the fate of Billy Pilgrim in Kurt Vonnegut's well-received novel *Slaughterhouse-5.*)

For those testing some of these options in Guin's story, the least-energy answer seems to be formulated by a stern Major, after rebel Bill Walden, the alter of Conrad Manz whose wife Bill has become involved with, is "dissipated":

> No, Bill Walden had been wrong, completely wrong, both about drugs and being split into two personalities. What one made up in pleasure through not taking drugs was more than lost in the suffering of conflict, frustration and hostility. And having an alter... meant, actually, *not being alone.*
>
> It was a pretty wonderful society he lived in, he realized. He wouldn't trade it for the kind Bill Walden had wanted. Nobody in his right mind would. (p. 204)

This seems even more comic-strip-simplified than most philosophers' thought-experiments about consciousness, but it might form the basis for an intriguing seminar... combining a grim Aldous Huxleyite smile with a cold shiver down the spine. Our real futures are likely to be far more disruptive or even desirable, once we are there, than this. At least to one variant of our user illusion consciousness.

* * *

What Is It Like To Have an IQ of 500?

1954 Poul Anderson, *Brain Wave*

What if the intelligence of every animal on Earth were to be doubled, tripled, enhanced fivefold beyond the current mean? What would life, and awareness, be like then?

When the young physicist Poul Anderson was 26, in 1953, the same year he married, he sold a truly remarkable two-part serial to the lowly sf magazine

Space Science Fiction. This magazine lasted just eight issues and went bust with only the first half of "The Escape" making it into print. The story was quickly published by Ballantine books, as *Brain Wave*, the following year (and perhaps had been sold as both serial and novel at the same time by Anderson's agent). Within another half decade it was snapped up by editor Anthony Boucher for the second excellent volume of *A Treasury of Great Science Fiction*. It has been dubbed, appropriately, and despite some serious weaknesses, "a masterpiece," by Larry Niven. In a 1997 *Locus* interview, Anderson somewhat uncertainly included the novel in a list of his five books most likely to be remembered. For our purposes, it is especially salient, since the very nature of human consciousness and intelligence is changed overnight while emotional immaturity and heightened existential terror warp the renovated minds in very old ways.

The solar system, in its great circuit of the Milky Way, passes beyond the boundary of a vast sweeping field of force reaching from the core of the galaxy. This field has been inhibiting electrophysiology in all animals for at least 65 million years, and this probably is what caused the extinction of the dinosaurs. Now the inhibitory effect is abruptly gone, and many of the fundamental constants established by science are shifted slightly, requiring massive recalibration of instruments. As a side effect, neuronal and synaptic transmission speeds up, together with other subtle changes that improve the effectiveness of memory, imagination, IQ. Not only does this instantly disrupt and revise human states of consciousness, but all the mammals cultivated for food and service start to understand how vile their lot is, and how easy it is to escape from their imprisoned fate.

The narrative, brutally restricted to 164 paperback pages, skips from one character and setting to another, but mostly focuses on a team of researchers whose New York Institute is funded by elderly philanthropist John Rossman. Notable among these brilliant mind workers is 33 year old Dr. Pete Corinth (IQ going in 160, swiftly soaring to 400 and above, whatever that means), beautiful Helge Arnulfsen who loves him chastely, bulky biologist Nathan Lewis, union boss Felix Mandelbaum, several spouses including Pete's pretty, nervy wife Sheila. This cast, despite Anderson's then-liberal attitudes, is strikingly male-centric.

As the world passes through the horrific violence and confusion of readjustment, Sheila is paralyzed by her inability to do anything beyond what is arranged for her by men. Ultimately she is relieved of the unsupportable pressures of enhanced consciousness by electroconvulsive treatment, which burns out parts of her brain and drops her IQ back to a comparatively moronic 150 points—where 100, by definition, used to be the population average. (In the real world, of course, ECT does not have this atrocious result, although it can interfere temporarily with memory formation.)

On Rossman's estate, an amiable, motivated moron (his own word) named Archie Brock looks after numerous stock animals, pigs, horses, sheep, chickens, aided by his smart dog Joe. As the inhibitory field is dispelled, Archie and Joe grow smarter, but so too do the resentful farm creatures. Horses trample on their harness, refuse to comply. One pig noses up the latch and leads the others in an escape into the woods. While Archie is still cleverer than any of them, the sheer bulk and number of the larger beasts make them formidable. With an IQ now in the former genius range, Archie manages to sustain the farm with the help of escaped chimpanzees and a tractable elephant, despite the collapse of food and supply markets and power service.

He is our Virgil in this super-bright new alarming world, his robust yet sweet-natured humanity holding us in the story even as the other characters become ever more alien, their speech reduced to inferences that have to be expanded for us by Anderson in what amount to sidebars. It is an interesting experience, but the transition to this new kind of consciousness is not easy to empathize with—or, where it is, to accept at face value. In China, a saint approaches a group of anti-communist insurgents. Despite the frigid snowy weather, he is thinly clad yet comfortable. He will teach them the secrets of self command now available to the expanded mind.

In a bow to traditional sf tropes, the supergeniuses swiftly develop a grand unified theory welding together general relativity and quantum theory. From this they determine that the speed of light actually isn't the cosmic limit, and construct a beautiful starship based on this new knowledge. Corinth and biologist Lewis are chosen to take it for a test flight to neighboring planets, then to the star Alpha Centauri. Predictably, they get caught up in the outer fringes of the inhibition field and, minds closing down, can no longer understand the nature of their radical new craft or how to steer their way clear.

After weeks in the daze of their former minds, they drift out of the field and

> their lives since Earth had left it seemed dreamlike. They could hardly imagine what they had been doing; they could not think and feel as those other selves had done. The chains of reasoning which had made the reorganization of the world and the building of the ship possible within months, had been too subtle and ramified for animal man to follow. (p. 116)

This resurgence of mind, Corinth understands, is "an unending struggle between instinct and intelligence, the involuntary rhythm of organism and the self-created patterns of consciousness. . . a life span of many centuries ought to be attainable" (pp. 117–18).

They find a star like the Sun, a planet like Earth, inhabitants not unlike the unaltered humans. They leave, find another, then nineteen in all, fourteen with intelligence but all marooned at the level induced by the inhibition field—without ever having encountered that field. Evidently evolution strands life at our station, unless it is blocked and forced by contest to supercharge its stifled brains. Only the freed superhumans of Earth are elevated in this way, perhaps, as some claim, by the choice of a God, or perhaps in the next step to their becoming gods. Anderson's narrative winds onward, sometimes compacted, often lyric and lovely. Brock, leader and custodian of a community of "morons," takes Sheila, with her damaged mind, as his new companion. Lewis visits, and tells Archie:

> For the first time, man will really be going somewhere; and I think that his new purpose will, over thousands and millions of years, embrace all life in the attainable universe... We will not be gods, or even guides. But we will, some of us, be givers of opportunity... No, we will not be embodied Fate; but perhaps we can be Luck, And even, it may be, Love. (p. 162)

That is surely a hopeful prospectus for consciousness in the cosmos, emergent and embracing rather than metaphysically separate and (somehow never explained by those who prefer this version) immaterial.

* * *

What Is It Like to Be in Love With a Crab?

1960, 1969. James White, "Countercharm."[2]

One of the most congenial of all science fiction galaxies was the invention of Northern Irish writer James White (1928–1999), whose Sector General stories and novels are set in a vast multi-species and pacifist deep space hospital. This immense and variegated gravity-controlled structure has facilities for almost all the known bioforms and ecologies of its interstellar catchment region, each species and environment coded with four letters marking the critical differences between them. Physiologically, humans are DBDGs, while the empathic Duwatti snails are EGCL, and the Crepellian many-limbed water breathers,

[2] *New Worlds* 100, Nov 1960, pp. 101-23; reprinted in *The Aliens Among Us*. Del Rey.

resembling octopuses, are AMSLs. There are numerous other sentient beings; VTXM are hive-minded beetles fed on raw radiation, while delicate low-gravity birds are MSVKs. These classifications are far from self-evident and rarely explained in detail, but their use provides a subtle sense of reality to the stories, a notional underpinning to a range of species differing not only in diet, breathable air, reproduction and so on but in styles of thinking: of consciousness, in fact.

Most of the Section General tales concern Dr. Peter Conway, a young human physician learning the ropes usually by being thrown head first into an alien setting and depending on his training, intelligence and empathy to carry him through. By the later tales, he has advanced to Diagnostician in Charge of Surgery. He quickly falls for voluptuous senior nurse Murchison, who later herself becomes a doctor and Conway's wife. His early infatuation is key to the solution of a typically startling problem addressed in this story.

Since there are numerous varieties of beings using the services of Sector General, and staff are called upon to deal with many that are distinctly different from themselves, a technology has been developed to store species-specific information and port it directly to a doctor's mind via an Educator tape. It will remain lodged there for long enough to allow treatment of whichever kind of patient needs help, before being deleted.

In "Countercharm," Conway is obliged to adopt the memories and cultural background of an ENLT life-form so he can instruct a visiting gathering of that species in a new surgical pancreatitis breakthrough, enabling them to return home with facility in this skill. (Why they can't download this information directly is not made clear.) These Melfans are a passionate, intensely emotional species; he finds that the being who made his ENLT tape was a "hellion where the females were concerned." Senior Physician Conway, lecturing the specialists in the new technique, is drawn against his will into this mind-set, finding himself increasing besotted with, although unrequited by, Dr. Senreth, female surgeon... crab.

> The object of his affection was one of the... six-legged, exo-skeletal and vaguely crab-like beings from Melf Four—and as the lecture proceeded his gaze was drawn to this entity more and more frequently, and became almost lascivious in its intensity. One half of Conway's mind—the sane, human half—kept insisting that getting all hot and bothered about an outsize crab was ridiculous, while the other half thought lovingly of that gorgeously marked carapace and generally felt like baying at the moon. (p. 2)

What's especially interesting, for a story originating in the British magazine *New Worlds* in 1960, is the candor of this description combined with a refreshing post-sexist and indeed post-speciesist attitude to the humans' space brethren and sistren. Not that White was advocating anything resembling bestiality (and that slur would be unfair in any case, given that all these mutual aliens are *people*, not animals). Conway's anxious concern was that he might make a fatal error during surgery, easy to do with these patients whose exoskeleton blocked access to the very delicate pancreas. If his second-hand obsession with the beauteous Senreth distracted him at a crucial moment, his patient might perish and his career would be over.

What's worse, his superiors refuse to heed his request to be replaced, or to erase his temporary Instructor programming. Finally a friend, Mannen, hints that he might have a solution, but cannot reveal its nature. Prepping for an exemplary operation on a suffering Melfan, Conway finds Nurse Murchison gowned and waiting. As the pressure to succeed intensifies, he is additionally distracted by Murchison's garb. "A nurse of her experience should have known better. And her belt was definitely too tight. The effect, in other circumstances, would have been distracting to say the least" (p. 19). He returns to his task "confused, excited and overstimulated in some odd fashion" (p. 20). Yet his hands are steady.

> He still regarded Senreth's mandibles as beautiful—hard, steady, wonderfully precise appendages which it was a joy to behold in operation. But when he touched one it felt like a warm, slightly damp log. . . (p. 20)

The operation is completed perfectly. Murchison, he understands abruptly, has been planted on him to offset his attraction to the crab with the countercharm of one of his own kind, "the simple answer to a complex psychological problem" (p. 21). He flings her around in a mad dance.

> A little wildly he said, "Murchison, I love you all to pieces. You'll never know why, but I've got to show my appreciation somehow. . . ."
>
> "Dr. Conway," said Murchison gently, temporarily ceasing her attempts to pull free. "I may never know but I can guess an awful lot. And I flatly refuse to catch someone on the rebound from a six-legged, female crustacean. . .!" (p. 21)

Of course we may now reproach this story for the jocular adolescent sexism of its solution, yet it remains charming, amusing, and an intriguing gesture at the correspondences and yet extreme dissimilarities in states of consciousness between individuals, even in the depths of space and with no evolutionary biases to anchor the artificial allure of accidental Instructor programming that might entice a luckless human to go where no human has gone before.

6

Entering the Mainstream

The words we think seem to hover in some insubstantial interface wherein we understand neither the origins of the symbol-signs that seem to express our desires nor the destinations wherein they lead to actions and accomplishments. This is why words and images seem so magical: they work without our knowing how or why.
Marvin Minsky, *1988, p. 196*

What Is It Like to Be a Ship?

1961–69 Anne McCaffrey, *The Ship Who Sang*

A prolific American science fiction and fantasy writer who sang, in 2005 Anne McCaffrey (1926–2011) was the third woman named as a Grand Master by the Science Fiction and Fantasy Writers of America. She had relocated with two of her children to Ireland in 1970 after divorce, taking advantage of that nation's exemption from income taxes for resident artists. A popular if stylistically pedestrian writer, she won both a Hugo award (for "Weyr Search," published in *Analog* in 1967) and a Nebula (for its *Analog* sequel "Dragonrider" in 1969). These launched one of her most admired series, about telepathically linked humans and teleporting dragons on a distant world. But an earlier series, first comprising shorter stories by McCaffrey alone and then novels co-written with collaborators Margaret Ball, Mercedes Lackey, S.M. Stirling, and Jody Lyn Nye, began in 1961 with "The Ship Who

D. Broderick, *Consciousness and Science Fiction*, Science and Fiction,
https://doi.org/10.1007/978-3-030-00599-3_6

Sang," in *The Magazine of Fantasy & Science Fiction*, and five others from *Analog* and *Galaxy*.

The central device in these works confronts many prejudices. Helva was born cruelly damaged, and saved from euthanasia by surgical removal of her "crabbed claws" and "clubbed feet," then by transition into a metal shell where her nervous system was adjusted to "manipulate mechanical extensions. As she matured, pituitary secretions manipulated, more and more neural synapses would be adjusted to operate other mechanisms that went into the maintenance and running of a space ship" (p. 1). Even in the twenty-first century, with cyborgs and implanted access to the Internet increasingly recognized as imminent realities, the proposition that redesigned babies might be transferred into titanium shells and trained as pilots remains dismaying to many. It is not so for Helva, who at age 16 is grafted into her starship, rejoices in her power and autonomy (although she has to buy herself and her ship back, as students now pay off steep degree loans). She replies sharply to a expression of sympathy that "it would be very difficult to give up hurtling through space and be content with *walking*" (p. 21). But Helva is not a robot. When her ambulant onboard male scout Jenna (a "brawn" such as most "brains" carry) gives up his life in a rescue operation, Helva is plunged into grief. She has sung with Jenna, whom she loved. Now, as Service aircraft

> dipped in tribute over the open grave, Helva found voice for her lonely farewell.
>
> Softly, barely audible at first, the strains of the ancient song of evening and requiem swelled to the final poignant measure until black space itself echoed back the sound of the song the ship sang. (p. 17)

This mournful Last Post, or Taps, closing the first story in the collection, is reminiscent of certain passages in Heinlein (such as the songs of Rhysling the blind space poet, or "Buck's Song" that wrenchingly closes "The Tale of the Adopted Daughter" in *Time Enough for Love*), which is perhaps one reason why McCaffrey received the Robert A. Heinlein award in 2007. It might seem insensitive to object that the sound of a funereal song can't be "echoed back" by black empty space. But it is necessary for McCaffrey to take such poetic liberties, because she has carefully denied this desolated brain in a titanium can the physiological opportunity to weep, and Helva regrets this loss. Despite her disabilities and their corrections, she has the consciousness of a young woman whose lover has perished *as he enters her (metallic) body*, yet also, in a tormenting unconscious link, to a miscarriage *from* her body.

This readjustment of the canonical trope of spaceship as womb serves to make Helva and others of her kind both familiar and creepily alien. We cannot forget the other major trope, sleek spaceship as penetrative penis, which is entirely subverted by Helva's indwelling presence *within* the classic thrusting shape of a spacecraft designed specifically to plunge through the burning heat of an atmosphere. It is surely no accident that the cover art of the 1969 Del Rey edition, by The Brothers Hildebrandt, shows against a black sky the disembodied face of a beautiful showgirl with sweeping blonde hair. In front of her ghostly presence a traditional pre-Apollo spacecraft rises on white flame from a ruined landscape, volcanoes jut, spurting gouts of lava, and a large ambivalent cone of rock in the foreground, somewhat emulating the shape of the ship, is pieced on one side by a great vulval vent, or perhaps a shrieking mouth. Of course McCaffrey had nothing to do with this representation, but the tales of Helva (one of which is set in just such a brutal erupting landscape) do evoke these doubled signifiers.

It is worth recalling that Helva's saga is not the first time the trope of a brain or even a severed head in a bottle has appeared in science fiction. An early instance was *Professor Dowell's Head* by Alexander Belyaev, in Russian in 1925, with English translation delayed until 1980. As George Zebrowski notes:

> It succeeds in being vigorous and unsilly (don't ask me how), even though it is dated. The power and zest of great pulp stories cannot be denied. Whenever I come across one, I feel that our junk today is neither stylish nor classy, or intelligent; there was better junk in the old days. (*F&SF*, "Books," April 1981, p. 60).

Nor was *The Ship Who Sang* the last to entertain this device. Heinlein's *I Will Fear No Evil* (1970), published just a year after McCaffrey's fix-up novel, explored the combination identity of Johann Smith, a billionaire of extreme age, whose brain is transplanted into the murdered body of his lovely young secretary. His mind emerges from this traumatic procedure, but somehow Eunice Branca's own consciousness continues to chat with him (from memories stored throughout the nerves and synapses of her trunk and limbs? from a postmortal archive?), and this doesn't seem to be a delusion. They remain distinct, but the sexist brain donor learns better by being a woman—and even falls in love, and into bed and pregnancy, with his oldest male friend.

In any event, the singing starship, like the singing swords of certain fantasies, is a memorable image of transformation, loss and recovery. Helva has messages for us to ponder as we consider the mysteries of consciousness and its source and immersion in the flesh. She does not do without the body, but her experience of the body is truncated and repurposed. Perhaps the same can be said of us, anchored to our smartphones and cars and computer interfaces. But at least, like Helva, we can enjoy the music of *Tristan und Isolde*, *Candide*, *Oklahoma*, and *Le Nozze di Figaro*, along with the atomic age singers, Birgit Nilsson, Bob Dylan, and Geraldine Todd, if not yet "the curious rhythmic progressions of the Venusians, Capellan visual chromatics, the sonic concerti of the Altairians and Reticulan croons" (p. 4). But there is always Taps, that "simple, poignant statement of end and beginning" (p. 177)—a farewell to consciousness, unless it proves to be, as Heinlein suspected, the opening into a larger realm of mind.

* * *

What Is It Like to Be a Fish?

1962 Brian W. Aldiss, "Shards"; 1962 "A Kind of Artistry"[1]

To begin at the end, thereby ruining any eager readerly anticipation of learning about the seeming aquatic aliens trapped in a large tank: these are named in the final line as "tunnies," a term for Atlantic bluefin tuna or Southern bluefin tuna. Both varieties are frequently eight feet long and weigh perhaps 550 lbs (the largest recorded Atlantic bluefin was twelve feet long, weighing in at 1500 lbs). And in the case of "Shards," two such monsters have been modified surgically in a military attempt to spy on actual aliens that landed in the North Atlantic, off Bermuda. Submarines found the deliberately destroyed remains of the spacecraft on the ocean bed. The aliens are already responsible for floods along the American and European seaboards and inundating the West Indies. It is clearly an alien invasion of the planet, even if it is restricted to the oceans.

The surgical modification is voluntary but nightmarishly cruel: in each case, one hemisphere of a human brain has been inserted into a tuna's nervous system. The hybrid has to make sense of both its new environment and the disorder of its own shattered consciousness, largely conveyed to us in soliloquy and dialogue wrought in a Joycean pastiche, the kind of narrative *tour de force* familiar from T.S. Eliot, the Dadaists, Samuel Beckett, and more recently Anthony Burgess. Aldiss was in his mid-30s when he played this exploratory game with narrative convention. At the outset we face this sort of thing

[1] Both these stories are collected in Aldiss's *The Airs of Earth*, New English Library, 1975, to which the cited page numbers refer.

(an approach that would expand in the psychedelic babble of his *Barefoot in the Head* and the cold coverage of *Report on Probability A*):

> The long liverish gouts of mental indigestion that were his thought processes would take over from his counting. . . . he named the present time Standard O' Clock [which] he pictures as a big Irish guardsman with moustaches sweeping round the roseate blankness of his face. Every so often, say on pay day or on passing-out parade, the Lance-Standard would chime, with pretty little cuckoos popping out of all orifices. As an additional touch of humor, Double A would make O'Clock's pendulum wag. (p. 82)

The Freudian gag/wag signals this mental territory as the boundary between bids at cognitive understanding and slightly transgressive unconscious play. The great fish imagines himself an amputee and only later excitedly discovers his tail—"His long strong tail induced him to make another experiment; no more nor less than the attempt to foster the illusion that the tail was real by pretending there was a portion of his brain capable of activating the tail. More easily done than thought." In the beginning, less secure in both epistemology and ontology, Double A masturbates his smooth body in the thick mud floor of his tank, yet shrinks in shame at naming what he does.

He calls for dark glasses and they are provided, then for further surgery to remove his troubling eyelids. At last he understand that these have been hallucinations, ways of accommodating to his horrendous new shape and circumstance. He discerns another of his kind, Gasm. They trade prankish jests, and Double A looks for his role in literary character conflict, "which pings right out of the middle register of the jolly old emotion chasuble. Amoebas, editors, and lovers are elements in that vast orchestra of classifiable objects to whom or for whom character conflict is ambrosia." Identity in Mudland is the first declaration of conflict:

> "What is your name?"
> "My name is Gasm."
> "Name some of the other names you might have been called instead."
> "I might have been called Plus or Shob or Fred or Shit or Droo or Pennyfeather or Harm."
> "And by what strange inheritance does it come about that you house your consciousness among the interstices of lungs, aorta, blood, corpuscles, follicles, sacroiliac, ribs and prebendary skull?"
> "Because I would walk erect if I could walk erect. . ."
> "Let us sing a sesquipedalian love-song in octogenarian voices.
> . . .*No constant factor in beauty is discernible.*
> *Although the road that evolution treads is not returnable. . .*

The bosom's lines are only signs
Of all the pectoral muscles' tussles
With a fairly constant factor, namely gravity.
They fell back into the mud, each lambasting his mate's nates.[2] (pp. 85–86)

This transition from garbled human consciousness into the piscatorial condition via a sort of imagined fetal jollity is startling but effective. The two scientists watching from above the tank detect speech patterns between the modified humans/fish, as they enter a realm of consciousness that is literally diluvian, threatening an even more alien foe. There is perhaps a hint, though, that the transformation might threaten humanity in the medium and especially the long term, as post-traumatically-stressed warriors coming home from patriotic carnage will sometimes turn against their own nation with violence that nobody, oddly enough, anticipated.

> Now the questions were no sooner asked than forgotten, for by a mutual miracle of understanding, Doublay and Gasm began to believe themselves to be fish. . .And then they began to dream about hunting down the alien invaders. . .. In the tank, in the twilight, the two giant tunnies swam restlessly back and forth, readying themselves for their mission. (p. 89)

* * *

In the same year, 1962, that "Shards" was published in *The Magazine of Fantasy & Science Fiction*, a somewhat related and beautifully-wrought story with an aquatic and species-uncertain tenor appeared in the same magazine. "A Kind of Artistry" has often been anthologized, and more than once the encomium of its title has been directed back, appropriately, by editors and critics at the story and its author. In the text itself, though, we encountered the reflection of an engineered slave: "We Parthenos will never understand the luxuries of a divided mind. Surely suffering as much as happiness is a kind of artistry?" (p. 32).

Near the dawn of what would be dubbed the New Wave, Aldiss's literary ambition found a luminous opportunity in this story of matriarchal domination, alien contact, and once again a species transformation more poignant than anything dreamed by H.G. Wells's Moreau.

In an interstellar community where humankind has modified itself in numerous inventive ways—the Earth-born Derek Flamifew Ende warmreads

[2] One small defect in this Monty Python merriment is Aldiss's apparent belief that "nates" rhymes with "mates," whereas this medical word for "buttocks" is actually pronounced ˈnā-ˌtēz.

messages from a distant star via a receptor bowl, like Joseph Smith poised over his scrying hat, detecting its signal via "the boscises of his forehead" (p. 13). These are presumably sensory snouts able to receive and transmit information, a tiring process, "as if the sensitive muscles of the countenance knew that they delivered up their tiny electrostatic charges to parsecs of vacuum and were appalled" (p. 14). Derek's Mistress or wife-mother, a century his senior and metaphorically a sort of queen bee, bears the same detectors. He has two hearts, again actually and metaphorically. Other people he meets are far more unusual in physical structure, and perhaps emotionally: a woman who flirts rather desperately with him is a "velour" with silky brown fur, while a major diplomat, an unglaat, carries on his back and head an elaborate antler system with six eyes. A common patois hides the variations in consciousness, in happiness and suffering: "each center of civilization bred new ways of thought, of feeling, of shape—of life" (p. 18).

Derek's Mistress is a scientist intent on unlocking the rigidities of species. In an early moment of bitter disagreement, Derek runs from her, and she seizes up a minicoypu—a small ratlike beaver—and flings it into a tank of water. "It turned into a fish and swam down into the depths" (p. 16). He loves her, cannot leave her, cannot stay with her for long. When he is called to investigate the dark world Festi XV, he goes at once, and finds there a vast entity that seems a self-engendered awareness built in the center of a fallen asteroid. "[The] thing possessed what might be termed volition, if not consciousness" (p. 20). The two extremely contrasted orders of mind engage each other, and Derek returns to communal gratitude, and loneliness. "Soon he would be back at Endehaaven. Though he would always be a failure in his dealings with other humans, there at least he knew every inch of his bleak allotted territory" (p. 31).

And there, with his Mistress refusing to greet him, he wanders among her experimental tanks, nets a fish and flings it into the air, where "it struggled into new form and flew away" (p. 31). Its cellular bondage is unlocked, but Derek's conditioning, his very consciousness, remains impenetrable. At last, viewing the sunlit fjord, in provoked fury, he strikes his wife-mother across the face. She goads him; he understands "for the first time her true nature." He throws himself over the cliff edge. "Even as his body hit the waters of the fjord, it began to change... A seal plunged into view, dived below the next wave, and swam toward the open sea over which already a freshening breeze blew" (p. 34).

* * *

What Is It Like to Be a Laminated Mouse Brain?

1963. Cordwainer Smith, "Think Blue, Count Two"

One of the most extraordinary minds attracted to science fictional mythoi, especially those of his own creation, Paul Myron Anthony Linebarger, PhD (1913–1966) was one of the pioneers of psychological warfare, a US Army colonel with CIA links, a godchild of Sun Yet-Sen, fluent in Chinese and Japanese, professor of Asiatic Studies at Johns Hopkins University. In his spare time, and dreams, he was bard of the Instrumentality of Mankind, an interstellar civilization tens or perhaps hundreds of thousands of years in the future. He published these under the curious name Cordwainer Smith. A smith is an artificer, and a cordwainer a shoemaker. The pseudonym catches something of the man's approach to fiction, plain if technically informed yet poetic, cadenced rather than high-flown yet magically strange. (He wrote other novels as Carmichael Smith and, translated from the Chinese for "the forest of incandescent bliss," Felix C. Forrest.)

This treasury of an improbable future history is collected in a sturdy 670 pp volume, *The Rediscovery of* Man, plus Smith's single Instrumentality novel, *Norstrilia* (pronounced Nor-stryle-yu, not Nor-strill-ee-yuh). That novel centers on the machinations and tribal ways of the immensely wealthy planet Old North Australia and its all but immortal inhabitants; Linebarger had spent time in Australia, which he dearly loved and to which he meant to move but a long illness killed him when he was just 53. Such ironies are frequent in his fiction as well. He was a cat lover, and many of his enhanced or uplifted Underpeople were transformed from feline or other animal stock.

The most memorable are C'Mell, a geisha-like "girlie-girl" who helps free her kind from the rule of humans, D'Joan, a dog person who re-enacts the drama of Joan of Arc, and T'ruth, an all but deathless little turtle girl who attends to The Mister and Owner Murray Madigan on the Storm Planet. Rather like Asimov's robots whose names start with "R-" the Underpeople are nominally marked by, for example, C' for cat-stock and T' for turtle ancestors. The E-telikeli is a great-winged Eagle shaman who plots this revolution. But humans, too, bear odd names. C'Mell's love-object is the old Lord Jestacost, but other Lords and Ladies of the Instrumentality include Lord Luvaduck, Lord William Not-from-here, Commissioner Redlady, Lord Sto-Odin (the number 101 in Russian) and the long-dead but uploaded ancient Lady Panc Ashash.

The sheer abundance of these names and places generates an abiding sense of reality even as they denote situations closer to fairy story. It is apparent that Linebarger had much to tell us about varieties of consciousness and its embeddedness in flesh of human or other kinds, not to omit minds uploaded into machines, and sarcastic robots. In the early story "Think Blue, Count Two," an unutterably beautiful 15 year old girl, or perhaps younger, is assigned by space authorities to travel in a starship propelled by light-sails, drawing after it a thread of frozen immigrant bodies that will not be defrosted for some 450 years. Their destination is Wereld Schemering, a colony where too many people are ugly; this flight comprises only beautiful people, whose consciousness and memory are only of a gilded life free of disease or discomfort.

The child is named after the last two numbers of her full identification: she is Veesey-koosey, with a Daughter Potential of 999.999, "meaning that any normal adult of either sex could *and would* accept her as a daughter after a few minutes of relationship" (p. 131). She is prepared for the journey by a psychological guard named Tiga-belas (Indonesian for 13), who builds a device based on a laminated mouse brain and a mental key ("Think blue, count two, and look for a red shoe") to protect her against harm if the ship fails and degenerates into ruinous behavioral horror. The horror is unspecified, being unspeakably vile: on the drifting starship *Old Twenty-two*, the sailor died or lost command. The frozen passengers were awakened, and

> did not get on well with one another. Or else they got on too horribly well, in the wrong way. Out between the stars, encased only by a frail limited cabin, they had invented new crimes and committed them upon each other—crimes which a million years of earth's old wickedness had never brought to the surface of man before." (p. 130)

Veesey-koosey's comrades, when she wakes 326 years later to an emergency, are a very handsome man, Treece, and a shockingly ugly one, Talatashar, who has been ruined by freezer spoiling. Both men are quickly and innately rapacious, Talatashar more explicitly. Smith's exploration of this transition and its consequences approach Alice Sheldon's feminist insights (she wrote as James Tiptree, Jr.)—or is it, perhaps, just a sociobiological version of Original Sin? Talatashar slaps her hard, to her disbelief, then rails in Luciferian *non serviam* mode:

> "I'm going to do what I please. What *I* please. Do you understand? . . . I don't want rights!" he shouted at her. "I don't want what's mine. I don't want to do right. Do you think I haven't heard the two of you, night after night, making soft loving sounds when the cabin has gone dark." . . . Talatashar groaned. The history of man was in his groan—the anger at life, which promises so much and gives so little, and despair about time, which tricks man while it shapes him.

> . . . "I don't hate girls," he blazed, "I hate me. . . . You're not a person. Girls aren't people. They are soft and pretty and cute and cuddly and warm, but they have no feelings. . . . I always knew that girls weren't people. They're something like robots. They have all the power in the world and none of the worry. Men have to obey, men have to beg, men have to suffer, because they are built to suffer and to be sorry and to obey." (pp. 144–45).

Expecting dismemberment and death, Veesy triggers her key. Immediately,

> a huge voice roared at them from the control board.
> "Emergency, top emergency! People! People out of repair!"

The mechanism of the ship reaches into the mechanism of the damaged man, and a beautiful woman enters from a storage room, wearing an old fashioned garment. In a peremptory tone, she says "Tal, be a good boy. You've been bad."

She is, of course, his dead mother, and instructs him to "take care of the little girl... If you don't, you'll break your mother's heart, you will ruin your mother's life... just like your father did."

This is darkly funny, but apparently does the trick. When the ship arrives at Wereld Schemering, and Tal's face is repaired, the two men speak kindly to her, apologize, even arrange to meet again. A pattern of utopian consciousness has been interrupted by horror and despair and the emergence of hidden unconscious drives lost in prehistory, but now the smooth flow is regained. In Cordwainer Smith's universe it will not last; the temporary melioration is far too slick. Men and woman will have to engage with the rediscovery of pain and illness and conflict, and the freeing of the Underpeople—by themselves and some well-disposed humans—and perhaps the robots as well. It is a great vast journey, and the tragedy is that Linebarger had scarcely begun its delivery.

* * *

What Is It Like to Be a Hieros Gamos?

1968 Josephine Saxton, "The Consciousness Machine"

Half a century ago, at the dawning of the age of, well, the New Age, of psychedelia and questing after the numinous and the imminent first flight of humans to the Moon, Josephine Mary Howard published a very strange oracular parable in *The Magazine of Fantasy & Science Fiction*. The June 1968 issue's main feature and cover-art novella, it was reworked into a slender 1969 novel titled, mysteriously, *The Hieros Gamos of Sam and An Smith*, as a Curtis paperback. (A "Hieros gamos" is an alchemical sacred/magical marriage or conjunction, highly significant to Jungians.)

Howard, or Saxton, was born in Yorkshire, UK, in 1935, and for years was an active adherent in "The Work," a form of practice devised by the Armenian mystic George Ivanovich Gurdjieff intended to awaken people from their robotic automatism and attain a higher consciousness.

This yearning for the expression in life of humankind's maximal potential finds a kind of representation in "The Consciousness Machine," where a training device called the WAWWAR—the Who And What We Are Room, its abbreviation evoking the howls of an infant—replaces standard psychotherapy by resolving psychoses in three weeks. Zona Gambier is a married operator of this still experimental device, employed by its inventor, Dr. Owenvaun. She attempts to break through the resistance of a louche and secretly drug-taking rapist, Thurston Maxwell, whose unconscious is purportedly probed by the machine and expressed in the form of a visual narrative that Maxwell and Zona watch together. She is secured against any physical depredations by a "protection circuit" that will knock him out instantly.

Unexpectedly, Maxwell fails to recover during this Jungian dramatization of what are supposedly his inner drivers, and it swiftly becomes obvious to the reader that these emblematic episodes are being plucked from the depths of Zona's consciousness rather than her charge's. A young boy drifting aimlessly through an arid, unpopulated world, stumbling upon a dead woman who has just given birth to a baby girl, is perhaps a kind of animus archetype conveying somewhat Tarot-card messages that Zona fails to interpret. In dream-like leaps of place and time, the boy George takes care of the baby, then explores the world with her when he is 21 and she is 7, creates a mandala in sand when a ferocious dog menaces them, undergoes a shock when she turns 14 and bleeds for the first time, allows her the name Beryl for the semi-precious jewel, but finally becomes himself Sam (for Samson, after she cuts his hair to force his maturation into manhood) and she is An (for Anastasia, "cherishing," "nourishing"). They make love after she "empt-[ies] her mind of thoughts, allowing them to slide from her, so that all her attention was centered on sensation and emotion" (p. 34).

When Zona is finally brought to grasp the meanings of these fantasias (within what John Clute calls "the Cave of the Jungian Shadow"[3]), she is first

[3] John Clute, "Josephine Saxton," *Encyclopedia of Fantasy*, online at http://sf-encyclopedia.uk/fe.php?nm=saxton_josephine

"exhilarated, joyful, and extremely strong," but fears being overwhelmed. Dr. Owenvaun reassures her, drawing on a concept from Aldous Huxley: "You are swept up by too much that is new. Too much reality all at once can be very frightening... but then your safety valves will close down, to protect you" (p. 44).

Earlier, testing the device himself, the experimenter had "clearly reached a higher state of being." Yet he was unable to warn Zona in advance of its likely effects, "for the conscious thought of a new, advanced state of being confounds the material from the unconscious." The events of the psychodrama she projected, supposing they came from the former madman Maxwell (now cured, luckily) were a cavalcade of symbols Dr. Owenvaun had already found in his own unconscious as well: "very few of them archaic or completely archetypal... with the main exception of the alchemical symbols, which remain true symbols for the philosopher's stone. Even if one calls it higher consciousness!" (p. 44).

In some respects, this novella is as pure a transduction from psychoanalytic constructs into science fiction, with an admixture of magical esotericism, as one might hope to find, if that is the sort of thing one hopes to find. The mysteries of the symbolic voyage are left entirely unexplained in the slightly lengthened novel version, which loses the WAWWAR, the doctor, Zona and Thurston. For the purest and zaniest version of Saxton's metaphysical drama, therefore, go directly to *The Hieros Gamos of Sam and An Smith*. But the *F&SF* science fictional adaptation (as I think it might be, after *Sam and An* was deemed too difficult by half for most sf readers to swallow dry) has its curious merits.

* * *

What Is It Like to Be the Absolute in a Can?

1969 Philip K. Dick, *Ubik*

In 1922, the Czech playwright Karel Čapek published a book with the bizarrely implausible title *The Absolute at Large*. (His brother coined the term "robot" for machines like humans, in 1920, and Čapek borrowed this for his play *R.U.R.*, the first story with an explicitly robotic cast—although these days the beings would be dubbed "androids.") Use of what amounts to matter-antimatter in mutual annihilation, for fabulous amounts of free power, releases as well a kind of immanent divine or sacred substance that pervades the planet and instigates horrendous global conflict.

Half a century later, the brilliant and grouchy Polish polymath Stanislaw Lem anatomized Philip K. Dick's hallucinatory novel *Ubik* by reference to a somewhat similar notion: the Absolute, with the powers of a somewhat constrained deity, is available as Ubik in an aerosol spray can or other container.[4] Ubiquitous in the future of the novel (if any of this can be trusted ontologically, or even epistemologically, since no description, belief, assertion or event in the world of the book can be relied upon), this ultimately mysterious substance is, according to Its own declaration, God, or the nearest affordable duplicate. Lem was not just being whimsical. In the novel's final pages, the pervasive stuff announces, as if it were the Tao personalized with a touch of Lewis Carroll, that

> I am Ubik. Before the universe was, I am. I made the suns. I made the worlds. I created the lives and the places they inhabit; I move them here, I put them there. They go as I say, they do as I tell them. I am the word and my name is never spoken, the name which no one knows. I am called Ubik, but that is not my name. I am. I shall always be. (p. 190)

This purported announcement by the creator, sustainer and organizer of everything might seem minatory, but actually Dick's manic novel is peppered with hilarious advertisements for Ubik, undercutting the gravity of this voice from a burning bush (so to speak; we never learn its source for certain). Chapters are headed with cheesy ads for this miracle substance, always followed by usage warnings (intended, of course, to be ignored):

> Instant Ubik has all the fresh flavor of just-brewed coffee. Your husband will say, Christ, Sally, I used to think your coffee was only so-so. But now, wow! Safe when taken as directed. (p. 21)

> Perk up pouting household surface with new miracle Ubik, the easy-to-apply, extra-shiny nonstick plastic coating. Entirely harmless if used as directed. Saves endless scrubbing, glides you right out of the kitchen! (p. 70)

> Lift your arms and be all at once curvier! New extra-gentle Ubik bra and long line Ubik special bra means, Lift your arms and be all at once curvier! Supplies firm, relaxing support to bosom all day long when fitted as directed. (p. 152)

[4] Stanislaw Lem, "Philip K. Dick: A Visionary Among the Charlatans," trans. Robert Abernathy, "Science Fiction Studies," # 5 Volume 2, Part 1, March 1975, https://www.depauw.edu/sfs/backissues/5/lem5art.htm

This splendidly charming parody of banal mass market ad copy, with that stinging little "Christ, Sally!" to remind us that we're not in 1952 (or even *Mad Men*'s 1960) any more. No, this is 1992, the drastically unlikely future presented in Phil Dick's actual 1969, although far closer to reality in this dawn of the Internet of Things. The world has been convulsed by market embrace of psychic phenomena such as telepathy and *anti*telepathy,[5] even limited control over time, allowing the past to be altered. The dead survive in cold-pac moratorium suspension, frozen but able to communicate with the living via a sort of Bluetooth until their energy is at last exhausted. Low grade but intensely irritating AIs demand payment to open doors or refrigerators, and threaten legal action if any attempt is made to strong-arm the appliances. Space travel is well established.

As usual in Dick's fiction, a cast comprises a wealthy but paternal boss, Leo Runciter, his wife Ella (currently dead, but employed by his psychic snooping agency, Runciter Associates), and lowly if competent workers such as Joe Chip, a computer operator whose very name embodies his role in the pre-cyberpunk future of the novel. Against these are ranged anti-psis such as Hollis, whose company is apparently responsible for killing Runciter's key operators and other staff. In a business meeting on the Moon, an explosion slays either everyone from the team except Runciter, who flees back to Earth, or only Runciter, whose corpse is returned for cold-pac treatment, or all or none of them.

The convergence here of epistemological uncertainty (what basis do any of them have for well-founded knowledge?) and ontology (what kind of reality *is* this mad universe, anyway?) is a quintessence of Phil Dick's increasingly mystic/Gothic presentation of the universe. This obsessive pursuit of what might be called the Absolute, in another sense, has been captured at excruciating length and variety in Dick's *Exegesis*, a sort of graphomanic philosophical journal published after his early death at 53, in 1982.

In a sense, the raw plot is not especially important in this case, or indeed in much of Dick's science fiction. Since the intention is to display unnerving and indefinite events, pulling the reader along by a chain that might at any moment turn into a necklace of roses, a bright beam of light, or a first century Christian fish icon, the causal absurdities and loose ends can be treated as part of the (serious) game of consciousness at the end or edge of its tether. Famously, in an earlier novel, *Time Out of Joint* (1959), the absurdly named protagonist Ragle Gumm sees a soft-drink stand—the ice chest of bottles, hot

[5] I dealt with the psi component often found in Dick's fiction in my *Psience Fiction* (2018, chapter 31), so will not repeat that here.

dog broiler, mustard jars, ice cream bins—fall into bits that go out of existence along with its proprietor:

> In its place was a slip of paper. He reached out his hand and took hold of the slip of paper. On it was printing, block letters.
> SOFT-DRINK STAND

In this case, the puzzle is resolvable: Gumm has been hypnotized and psychologically manipulated to believe that he inhabits the mid 1950s, earning a living by guessing where the Little Man will turn up in tomorrow's newspaper chart. In fact, he is gifted with precognition, and is foreseeing lethal missiles aimed at Earth's cities. The terrifying pressure is more than he can tolerate without this charade.

But in *Ubik*, there really does seem to be a kind of demiurge in control of everything. Does it function by manipulating the conscious experiences and unconscious urges of Runciter, his wife, Joe Chip, the rest of them? It is not clear, and *it can never be clear*. Chip finds messages from Runciter, after his death, and his face printed on bankbills. Runciter replicates this experience, except that it is Chip's face he finds on coins. We can't know. Perhaps the world, decaying into entropy and snatched back for a brief time by the application of Ubik, is just *like* this. Perhaps reality is, after all, endlessly decaying into dust and rubble, in a cosmogony of negative panpsychism. Alas, it is too late for us to check back with Philip K. Dick—unless, who knows? Perhaps he subsists somewhere in chilly half-life dreaming us all up, in the Beloved Brethren Moratorium (p. 8) or even The Bonds of Erotic Polymorphic Experience (p. 7).

7

Radically Different Minds

. . .one second before the solution popped into their heads, the visual cortex at the back of the head. . . briefly switched itself to a kind of "offline" state. The explanation for this, known as an "alpha blink," is that visual cortex shuts off incoming information for just long enough to allow the solution to a problem to pop through. . .. It take visual information out of the equation and lets the rest of the brain take that share of the available thinking power.
Caroline Williams, *2017, p. 124*

What Is It Like to Wake Up in Someone Else's Body?

1975 Lee Harding, *A World of Shadows*

Photographer and writer Lee Harding (1937–) was the first Australian sf fan to break into professional print at the start of the 1960s, with stories sold to John Carnell's British magazines *New Worlds* and *Science Fantasy*. His first novel, *A World of Shadows*, came out when he was 38, from the UK publisher Robert Hale, who apparently made some significant unauthorized cuts and other changes that weakened the published text, but the original version has never been released. Despite these depredations, the book has an unearthly tone of horror and grief that conveys the dismaying discovery that a deep space explorer's consciousness has been transplanted into the body of his crewmate while his own body is now comatose.

Humans have learned how to travel faster than light to other star systems by entering second-order space, but find in that realm frightening ghostly

D. Broderick, *Consciousness and Science Fiction*, Science and Fiction,
https://doi.org/10.1007/978-3-030-00599-3_7

creatures dubbed Shadows. These beings possess the ability to enter human bodies and minds. Once arrived at new interesting planetary systems, large exploratory vessels release small scout ships, each carrying two crew. Approaching the Canopus system, Stephen Chandler, an unprepossessing scholar, and handsome Richard Ashby, machine wizard, are trapped in space[2]. Waking from stasis, Chandler is terrified to discover that he is now in Ashby's body, a prankish or perhaps fact-finding intrusion by the higher-dimensional Shadows. The two and their ship are fetched home by the Shadows, 180 light years to Survey headquarters: "Without any fuss whatsoever, the *Polaris 3* reappeared in the sky a scant fifteen hundred miles above the northern Pacific ocean of Earth" (p. 29).

Chandler's wife Laura is naturally devastated, and watched closely by Chris Nolan, close friend of the Chandlers, an "alien psychologist" earlier victimized by Shadows but not as gravely:

> The atmosphere in the room was thickening with menace. His eyesight wavered and it grew impossible for him to see clearly. Something cold touched the perimeter of his mind and he knew at last that he was no longer alone… [The] familiar chill crept through his limbs and fastened his hands to the sides of his chair. He kept still. They converged upon him like trailing wisps of oily black smoke scurrying around the periphery of his distorted vision. (p. 11)

So far, so Gothic—with a hint of Cordwainer Smith's Space[3] and its terrifying indigenous menace, the Rats or Dragons.[1] But the book, which declares itself an "ontological thriller,"[2] lives up to that terminology, and opens fresh paths in our quest for the nature of consciousness as portrayed in science fiction.

Officialdom, of course, doubts that the body of Ashby now truly instantiates the mind of Chandler, taking as their default opinion that the new "Chandler" is a fake, compiled by the aliens for inscrutable purposes. Laura and Nolan become convinced that this is simple institutional paranoia, that the consciousness swap is real and persistent. Slowly Laura's love for her second husband returns and strengthens (she was widowed young, when her first spouse died in a vehicle accident). Luckily for the story, Ashby was not married; a strikingly handsome man, he played the field when he was not obsessed by machines. So there is no alternative wife to battle Laura for Chandler/Ashby's affections.

---

[1] In Smith's "The Game of Rat and Dragon," October 1955, *Galaxy*.

[2] This is the term employed on the inside jacket flap blurb, but was evidently proposed by Harding.

Nolan, once himself possessed briefly by Shadows, has specialized in their study, graduating as an "alienist" (a nice appropriation of the late nineteenth century term for a psychiatrist). He has previously disclosed to the couple, before Chandler's mind-switch, that the Shadows

> were disturbing the general consensus of reality. How one sometimes saw strange shapes moving across the sky... At odd moments the sky would seem to become translucent and strange figures could be seen moving behind it, as though it were a piece of smoked glass. Was it only imagination, or was their familiar continuum undergoing a subtle change? (p. 102)

It seems clear that these transformations of reality are not created or directed by human intention. Probably human consciousness is being manipulated by the aliens. Simple folk, Nolan explains, are relatively immune to their infestation. Clever complex minds are far more vulnerable. And Chris Nolan's wife Eleanor, deeply depressed, had killed herself. These bad memories return as the three finish an excellent dinner prepared by Chandler in a first attempt to make himself known again to his wife. In a nine page mental battle, something of a *tour de force* akin to Colin Wilson's similar scene in the Lovecraftian *The Mind Parasites* (1967),[3] Nolan and the couple are flung into a nightmarish virtual world, only to find after they are released that the Shadows have secreted inside each mind "a dense coal of darkness... an observant parasite keeping watch over their thoughts..." (pp. 113–14).

The final, grievous ontological trap remains to be sprung. Once husband and wife are reconciled finally to their altered fate, joined at last in orgasm, the Shadows betray them yet again. Chandler's consciousness is reawakened in his original body, thousands of miles away. Nolan understands, finally: with Ashby's suppressed identity set free, each man is once again conscious in his own body:

> "They took a duplicate they had made of your husband's personality and superimposed it over the quiescent mind of Richard Ashby. The man we saw was *not* Stephen Chandler. We were duped, Laura: all of us. The man we thought was Stephen was only an ontological double, a psychic twin of the real person." (p. 149)

[3] We can be sure of this, because Wilson himself tells us in a Preface to *The Mind Parasites* that his scene of battle with mind parasites is a *tour de force*...

Laura, now pregnant with the seed of a man who never existed, "the personality that had grown and experienced [Nolan's] tedious interrogation" (p. 158), knows now that Nolan is her true devoted lover. And for Chris Nolan, alienist, widower, perhaps set free like Laura from the poisonous dark stars secreted within them, "life had opened up all around him like the petals of an enormous flower that had lived too long in the gloomy undergrowth of a rain forest" (pp. 159–60). Ontology, one might hope, recapitulates cosmogony. Except, probably, for poor bereft and *faux*-memory-wiped Ashby, stumbling in the darkness.

* * *

What Is It Like to Be a Leo?

1976 John Crowley, *Beasts*

A literary event that made up for science fiction's frequent slovenliness was John Crowley's second novel, *Beasts*. At once robust and exquisite, it is a medieval beast epic reborn for the close of the second millennium and the opening of the third, a fable of the dangerous and seductive recovery of pre-lapsarian authority and submission. It possesses the measured, mournful quality of a penitential tour of the Unicorn Tapestries.

Beasts is richly conceived, cleanly and powerfully written, as far above the tawdry usages of most paperback sf as T.H. White's *Once and Future King* is above the raw thrills of a franchise movie. (It owes a great deal to that masterly volume, and its British author's troubled, closeted life.) And it offers an answer to the question Wittgenstein thought he answered satisfactorily, but probably did not: What is it like to talk to a lion, or at least, in this case, a human–lion hybrid? And what is it like to *be* such a creature?—or, rather, such a person.

Consider Crowley's heraldic display, or perhaps Tarot deck:

Heavy-set Painter is the King of Beasts, literally a blend of human and lion but named for the panther, or puma, created by genetic engineering yet marked for killing in an anti-hybrid gene-ocide, like all his kind. Tiny, Machiavellian Reynard, the legal counselor, is a blend of human and fox (a fusion familiar, if only to scholars, from medieval lore). Their competitive relationship forms the armature of the novel. In this balkanized North America, Dr. Jarell Gregorious, metaphorically Isengrim the Wolf, is the doomed Director of the Northern Autonomy, and his teen son Sten is a future king among humans. Caddie is an indentured teenaged girl who forms a sexual attachment to Painter, the Sun God of the leos, and finally slays Reynard. Loren Casaubon is a tightly-reined hebephile ecologist and custodian of hawks, obsessed with Sten.[4]

The full cast is more ample than this, often drawn explicitly from the medieval beast stories, and their own histories constitute a sort of Key to All Mythologies. The group consciousness of each beast-human variety differs markedly from anything common to humans—but in a more concerted way than the wretched creations of Wells's Dr. Moreau. Bearing in mind, of course, as with all these literary manifestations of the alien and theriomorphic, that they are creations of regular human writers, however inflected by careful or sportive observation of non-human animals.

And the inverse is true, as Reynard reflects: "There was no way for [him] to conceive of himself except as men had conceived of foxes. He had, otherwise, no history... [only] the legends of foxhunters. It surprised him how well that character fitted his nature; or perhaps, then, he had invented his nature out of those tales" (p. 51). As, indeed, John Crowley (and later, we as readers) had done on his behalf, and that of all the other genetic jumbles.

What, then, is it like to be a lion, or a lion-human? Cassie, deserted by her lion lover, perceives but cannot express the *Weltbild*, the world-picture, of the leo pride, which lived, so to speak, forever, but unarticulably:

> the females and the children: they lived within each moment forever, till the next moment. They took the same joy in the sunrise, hunted and played and ate with the same single-minded purpose, as they had when Painter had been with them; and their grief, when they felt it, was limitless, with no admixture of hope or expectation... leos aren't like Painter, not most of them. *Painter has been wounded into consciousness*, his life is—a little bit—open to us, something shines through his being which is like what shines through ours, but the females and the children are dark. You'll never learn their story because they have no story. If you want to go among them, you have to give up your own story: be dark like they are. (*my emphasis*; pp 154–55)

Taking the rough definition of consciousness offered at the close of Chap. 2 [the state or condition of being aware of the world (and of the self, however torn or fragmented or task-dedicated) via the evidence from the senses and the degree of successful match of our composite portraits of these experiences, and

[4] A splendidly insightful and scholarly analysis of these elements and their transformation is found in *Snake's Hands: The Fiction of John Crowley* (2003) edited by Alice K. Turner and Michael Andre-Driussi.

our memories of them and, from those, the projections of what we can expect in the future] we have to suppose that the minds of other creatures, or even creature–human hybrids, will be somewhat or even extremely at variance with our own. In a somewhat different sense of light and darkness specified above by Cassie, Crowley reveals the difference between hawk and dog:

> There are bright senses and dark senses. The bright senses, sight and hearing, make a world patent and ordered, a world of reason, fragile but lucid. The dark senses, smell and taste and touch, create a world of felt wisdom, without a plot, unarticulated but certain. (p. 175)

The world of the hawk is bathed in luminous brightness, while the dog's poor eyesight fails to distinguish especially between night and day, inside or outside, but his nose brings him news of "a universe of odors mingled or precise, odors of distinct size and shape, not yet discrete, not discontinuous, always evolving" (p. 175). This is also (as we shall see in the next chapter) the world of Whitley Strieber's *Wolfen*, and their roaming, intelligent, ferocious packs (and, for all we know, of the UFO abductors Strieber claims spirited him away to their craft to perform abominable invasions or surgeries on his paralyzed flesh). And the female and child leos are in some way imprisoned in a world without the lucidity and fragility of a hawk's high vision, adventitious, *without a plot.* This circumstance applies as well to the overall shape of *Beasts*, which is flooded with incident, archetype, beauty and horror, but closes in narrative suspension, with no predetermined path mapped out ahead. This is apparent, and by no means accidental, in the final words:

> "Shall we begin?" Reynard said. (p. 184)

* * *

What Is It Like to Be a Wolf in Wolf's Clothing?

1978 Whitley Strieber, *The Wolfen*

The first novel by Strieber, who in less than a decade would be globally notorious as the rectally probed victim (or so he claimed) of alien anal analysts from another realm of being, is a deft blend of horror, police procedural, and naturalistic science fictional development. It is also a strong presentation of inhuman consciousness in the form of an unknown apex species under global threat by human overpopulation. Is Strieber's lycanthropic peril meant as a metaphor, or just as how the world would be like if such creatures actually exist and hungrily prowl the ghetto nights of big cities? Or both? Perhaps simply this: the answer to "What is it like to be something like a werewolf, without the transition to and from humanity?" And a disturbing answer it is.

Two alert, competent young cops in the Auto Squad are sent on night patrol in a reeking, rusting sprawl of broken cars, the Foundation Avenue Dump. They are attacked with ferocious, devastating speed by large beasts, disemboweled even as they try to draw their weapons, and torn apart, then eaten. Sparse evidence of the beasts' nature is later collected, including plaster casts of their paw-prints, and three other kinds of metaphor are swiftly deployed.

The first is ubiquitous, a general 1970s sexism inevitably directed without conscious intent at one of the two detectives assigned to the case, Detective Sergeant Becky Neff, of the Brooklyn Homicide Division.

The second is the self-serving lazy attitudes of many at the top of both the police force and New York politics, including the Chief of Police.

The third, perhaps as ubiquitous as the first, at least in novels, is widespread corruption of the police force itself: humans feeding on the scraps they can extort from the criminal human gangs who prey upon and metaphorically devour the citizenry as their mode of existence. One of these is Becky's cop husband Dick, a captain in Narcotics, who takes bribes to pay for his Parkinsonian father's private nursing care, an ameliorating circumstance that perhaps represents as well the species' familial concerns of the Wolfen predators.

Since the novel is written for humans rather than evolved wolves, most of the narrative is focalized through the viewpoints, terrors and determination of Becky Neff and her older, uncouth senior officer, Detective George Wilson. The chief Medical Examiner, Dr. Evans, studies the butchered corpses but reaches no explanation. Several experts decline to speak their suspicions aloud or are ridiculed for doing so. One of them, Dr. Carl Ferguson of the Museum of Natural History, compiles a composite model of the attacking animals' foot structure; to the extent that it is wolf-like, it must be a branch of the species evolved separately for at least 10,000 years. But it is these creatures that represent Strieber's most interesting dramatization, from the inside.

Section by section, interleaved between blocks of text evoking the humans' concerns, wolfen packs and individuals slide in the shadows and rich, data-filled stinks of the city environment. Pursuing George Wilson to his well-armored residence, they are conscious of the stench of his fear.

> It was pitiful to see one so sick and so full of fear... Such a one needed death, and the pack longed to take him not only because he was potentially dangerous but because he was in the condition of prey. He needed death, this one... (p. 91)

The Wolfen are aware not just of their own social entity but that of the humans. Killing the young policemen in the Dump had been a mistake. *Our* young, one of them reflects, must not kill *their* young (p. 62). This is the ancient wisdom of the wild, where packs chase down the wounded and the old, but it also reflects what they know collectively of human society, where the elderly and miserable drift to the slums and attain "the condition of prey." It is true that the Wolfen are unusually intelligent for hunting animals, but even more importantly they are also, as Wilson realizes, gifted with astonishingly developed sense acuity. The human equivalents are almost stifled into uselessness. So the Wolfen inhabit a drastically disjunct *Weltbild.*

This different physical and perceptual universe is given to the reader in an approximation of the wolves' own specialized sensorium and modes of communication:

> They turned the body of their brother on his back and ate him, crushing even his bones in their jaws, consuming every bit of him except a few tufts of fur. He was eaten out of necessity and respect. They would always remember him now, his brave death and good life. Each of them committed the taste of his flesh to precious memory. Afterward they howled, this howl expressing the idea that the dead are dead, and life continues. Then they stood in a circle, touching noses, their joy at being together breaking through all the grief and upset, and finally they opened their mouths and breathed their heavy air together, their hearts transported by their intimacy and nearness. (p. 203)

Meanwhile, pursuit of their foes, Becky and George, continues relentlessly, tension mounting. When finally the pack breaks into an apartment to savage them, gunfire proves sufficient, if only just, to preserve the wounded humans and kill the intelligent animals. It is revealing that the final two pages of the novel, an Epilogue, are devoted to a Wolfen, not a human, viewpoint. Both parents have been killed by weapons, but the youngsters escape. "They felt loss but not defeat. What burned in their hearts was not fear but defiance; hard, determined, unquenchable" (p. 252). They howl. It echoes across the great human city, and is answered on the wind from pack after hidden pack, announcing their "powerful sense of destiny." It speaks also, of course, to our human experience of consciousness, but with an alien flavor beyond our own: a tincture only science fiction can provide.

8

L'Être et le néant

The main problem with trying to explain consciousness in the computational theory of mind is that it is impossible.... In order to understand what's going on in consciousness, as in the universe, the key will be to understand that there is an area where your intuition is applicable and thus meaningful, and there is an area where it is inapplicable and thus misleading.
J. Storrs Hall, *2007, p. 279*

What Is It Like to Be an Immortal?

1983 Pamela Sargent, *The Golden Space*

This is the world beyond the Transition, which is not what science fiction now often calls the Singularity or the Spike but is surely part of the way there, heading into transcendence from the human state. The key change is development of optional deathlessness, plus endless healthy youth to preserve immortality from the Tithonus curse. That was the Greek myth of the Trojan prince who gained endless life but alas his goddess lover carelessly forgot to request rejuvenation. The result was an indefinitely extended senescence, a fate that seems to threaten many of us in this interim epoch before the true Transition.

Josepha Ryba is 300 years old, and recalls her attempted suicide at 14, when she was fat, unattractive and lonely. Now she looks like a beautiful 22 year old, and has lived alone by choice for decades. Into her privacy comes Merripen Allen, who wishes for her to provide genetic material for a venture into the creation of new humanity: neither female nor male but both, without easily aroused instincts

D. Broderick, *Consciousness and Science Fiction*, Science and Fiction,
https://doi.org/10.1007/978-3-030-00599-3_8

and emotions but with sharpened, accelerated rationality. She agrees, stifling doubt; her first mutant child is Teno. Her friend Warner has Nenum, who like the other children does not howl for attention but simply offers "a steady cry... calm, steady and calm... I don't even know whether Nenum is my son or daughter. Am I supposed to call my child 'it'?" (p. 25). In this early 1980s' future the singular plural has not yet been adopted for the non-binary gendered.

In a remarkable feat of imagination blended with an invitation to our involvement, Sargent carries us forward from that inaugural moment for hundreds, perhaps thousands of years. It is a narrative device that would be used later by David Mitchell and Kim Stanley Robinson: characters (or their souls) moving steadily and calmly and sometimes, abruptly, traumatically, into a tomorrow where the very nature of human consciousness is repeatedly disrupted or purified.

By the nature of this maneuver, different people are sometimes brought forward to helm sequences shedding light on these *longues durées*: the dual-gender children, their parents, mad or desperate ideologues who yearn for death and seek to impose it, even generations wrought by alternative genetic and social changes. Perhaps the most disturbing is a tribe created by their "god," the scientist Domingo. Merripen and an associate Karim find themselves seized by an unshaven band of crudely clad men bearing the silver wands that are the common weapon of this era. Domingo saves them, explaining that he is the deity of several villages commanded by his recorded messages. These odd people are startling in the way philosophical zombies are imagined to be:

> "Their minds are divided; each side of the brain is separate. You see, they are not conscious of themselves. When their right side directs their left side, they hear it as a voice directing them—my voice, or that of someone with authority over them. They do not know self-doubt, self-consciousness, depression, and other such advances our minds have made... They live out their lives and die, but they do not really know death, because they continue to hear the voices and see the images of those who are gone." (pp. 194–95)

As discussed elsewhere in this book, that is a version of Julian Jaynes' once-popular model of the ancient two-part brain/mind, now conjoined and unified (to some extent) by the corpus callosum neural cable between cerebral hemispheres. Domingo justifies his barbarous experiment: "Can't you see?... There is no evil here. There is no sin, only innocence. It is a paradise, in a way. We ourselves might have risen from that state, or fallen from it." (p. 195)

In the most haunting section of the novel, the closing pages, wounded Domingo emerges from hibernation to find himself ill, probably dying. He stumbled into the caring hands of the descendants of his bicameral experiments. As he sleeps again inside a "transparent carapace," healing, a woman and two children tell or listen to tales of the Guardians, now long gone from Earth along with many vanished species. We know that the grown genius children of Teno and the other dual-sex not-quite-superhumans have passed to the asteroids at the edge of the solar system, carried there by a newly developed teleportation gate, and beyond there to the stars. Science and technology have long since devised the basic appliances and tools of the post-Transition, pre-Singularity communities. One is the materializer, providing all that's needed for nutrition and comfort (and copied jewelry or other art). Another is the transformer that dismantles the atoms of a person's body and beams them to a distant place for reassembly. What is left behind after most of humankind has gone to the stars is an aggregation of "dead worlds."

So are the Guardians just garbled or imagined tales told and embraced by the superstitious? "Life was once very hard," the woman tells her charges, "and people needed to believe in something... actually all we were perceiving was something inside our minds." But, she adds consolingly (and, as we are all but certain in the world of this novel, correctly):

> "...it's possible that the Guardians might have been real. We know that others, a very long time ago, changed themselves and lived in what we call the dead worlds. Why couldn't others have chosen to shed their bodies and transform themselves into something immaterial... Perhaps they were the ones who created us so long ago..." (pp. 245–46)

As indeed they were. And by the blind miracle of evolutionary pressure and drifting competing genes, the children of the bicameral zombies have awoken into true consciousness, again, finally, and sit to await the waking, all unknowing, of their creator Domingo.

* * *

What Is It Like to Be a Time Traveler?

1982–97 Spider Robinson, *The Lifehouse Trilogy*, compilation of *Mindkiller*, *Time Pressure*, *Lifehouse*

At the outset, I warned that this book is no respecter of *spoilers*, explicit revelations of plot arcs and character secrets, when such revelations are needed to make a point about treatments of consciousness in sf. That is especially true in the case of Spider Robinson's three volume ensemble known by the title of the third book, *Lifehouse* (a kind of pun on "lighthouse," a sturdy structure that stands between land and sea casting bright notice of perilous rock-torn seas, or in this case between life and death or memory and forgetting).

The key device or *novum* of this helter-skelter and emotionally involving if thrillerish and ingenious sequence is—(really, avert your eyes if you haven't read these books yet):

In the near future, via a conspiracy of scientists, humankind is, in effect, redeemed via manipulation of the brain's architecture, memories and drives: the consciousness of each human being. When retrocausal time travel is developed, this technology is the basis for an immense program of reclamation of the dead, all the way back, and their/our rebirth in a range of hive mind utopias.

* * *

The opening movement, *Mindkiller*, elaborates one of Spider's early successes, the 1979 *Omni* short story "God is an Iron" (that is, the deity specializes in bitter irony). In 1999, a skilled thief breaks into a young woman's residence and finds her near death with a wire plugged into a socket in her brain. This is the ultimate addiction, wireheading, where modules of the brain's reward system are triggered into ceaseless firing, overwhelming all other urges (such as eating, drinking, defecating in the approved manner, etc). Joe finds himself trying to salvage this woman, Karen, which he manages at the cost of a broken nose and copious blood. (It doesn't pay to interrupt a wirehead, no matter how emaciated.) We learn that Joe has no known surname; indeed he has lost most of his memory. What he does have is a fabulously comfortable den protected by advanced failsafes and security devices, gifted to him by an old dude named Fader, a criminal, it seems, with a remarkable ability to fade from pursuit. Fader, of course, turns out to be Joe's mnemonic thief.

Before we meet these two, though, six years earlier English professor and former soldier in the Africa war Norman Kent in Halifax, Canada, stands on a bridge in a brutally cold night readying himself for death. At the moment he is about to plunge into the water far below, an unkempt fellow grabs him and hauls him back. No guardian angel, this fellow is also a thief. Norman thwarts him in a vivid, gratifying scene, taking out his portable possessions and wallet—and hurling all of it over his shoulder into the water. It is an instant of reprieve. But Norman has motive for his suicidal gloom. His lover had left him for an unworthy replacement, and his beloved sister Madeleine has vanished.

Joe is (spoiler! spoiler!) Norman. With Karen, he learns that Fader is in fact the French neuroscientist Jacques LeBlanc who invented a device to erase selective portions of a victim's memory. Enraged, Joe and Karen track him down in order to kill him for this vile mind rape—but Jacques persuades them otherwise, in a remarkable *coup de théâtre* in which he tells them of his plan to strip power from the cruel, uber-rich, uber-powerful masters of the world and build a brighter future for humankind.

* * *

The narrative begins again, but decades earlier, in *Time Pressure,* set in Nova Scotia, Canada, in 1973. Narrator Sam is a rather misanthropic hippie musician in a hippie house some miles from a really hippie commune of vegetarian sexually various men and women. On a beastly cold night, Sam finds a large humming "globe of soft blue light... two or three meters off the ground.... The humming sound reached a crescendo, a crazy chord full of anguish and hope" (p. 13). The egg vanishes, and a naked dark entirely hairless woman is left aloft in its place. On her bald scalp she wears "a gold headband, thin and intricately worked" (p. 14) riding high on her skull. She is a time traveler, and a telepath, and Sam rather grudgingly and painfully hauls her inside to the warmth.

Naturally Sam falls for her, and after they make rapturous love she goes politely about the neighborhood in a polyamorous way, spreading illumination and preparing the path for... something wonderful. (Spoiler! spoiler!) Rachel is a kind of construct, blending the future consciousnesses of LeBlanc and his wife Madeleine, and the latter's brother Norman/Joe and his wife Karyn. Tomorrow's humanity is now The Mind. "To join The Mind you did not have to lose your ego, your identity or free will. You could leave The Mind and restore the walls around your own personal mind as easily as switching off a phone—that being in fact how it was done..." (p. 220). In the enormous reclamation of humanity lost to death in the far past, The Mind's agents would implant in every infant's spinal fluid "a tiny and fantastically complex descendent of a microchip which would copy every memory that brain formed—and when triggered by death trauma, would transmit that copy to the nearest buried 'bubble' for storage and future recovery" (p. 227).

* * *

In *Lifehouse*, in 1995 Vancouver, Canada, one of these storage bubbles is all but dug up by a mook looking to hide away his latest stolen hoard. He is witnessed by a far more engaging criminal, June Bellamy, a clever 33 year old scammer exercising in the forest reserve. She is startled when the sleazebag abruptly goes into orgasm, pants still on, hands off. Shortly after, leaving an unfinished message on her lover's phone and touching the mook's abandoned shovel, she too comes in her pants. And then loses her memory of the event.

She and her confidence man don't know what's happened, but we do. This is surely the work of future time travelers under the direction of memory thief LeBlanc and other salvific conspirators. And we have already met two of them, at the very start of the novel—an extremely old but well maintained couple, Myrna and Johnson, married nearly a millennium, whose Tantra is so all-encompassing that they miss June's accidental messing with the bubble and so "the whole universe very nearly ceased to ever have existed" (p. 4). Spider Robinson, like many sf writers, plays for the biggest stakes.

Meanwhile, two obese science fiction fans and computer adepts, Wally and Moira, find themselves on the night of Halloween in much the same situation as Sam when Rachel materialized—but the details are significantly different. Under a bright moon, a flaring magnesium-bright light flashes inside their property's boundaries. A bald naked man, "well muscled and trim," perhaps 25, a younger, taller *Star Trek Next Generation* Captain Picard, is hunched in a patch of scorched grass. He babbles in a kind of understandable cant, establishes the year, and is dismayed—more so when he learns that his hosts are sf fans, and faints. What else could he be but a time traveler?

Paul Throtmanian is, however, no such thing. June's lover, he is an ace scammer, and this will be his finest achievement, once he gulls the fans (sworn to secrecy lest, he tells them, they ruin the future) for $98,000. In doing so, he effects "*the first new con in at least a hundred years*" (p. 57). When the real time travelers show up, desperate to short-circuit the paradox implicit in releasing news of *fake* time travelers, which must inevitably blow their own cover and *really* ruin the epochal plans of The Mind, the story gets intensively recomplicated. Terror ensues, and mirth; also tension, apprehension and dissension.

* * *

What do we learn about consciousness from these enjoyable and highly improbable novels? That the individual self is vulnerable and uncertain, open to obliteration or modification by technology. That seeking ultimate pleasure as a goal, exemplified by brain-jolted orgasm, is going to be a consumer purchase or inflicted injury, trading consciousness for bliss of a mindless kind. That the same technology might open the way to mind-sharing of a truly telepathic sort, to mutual understanding and empathy, to an abiding concern for others of one's species even to the point of saving the dead in a kind of godless hitech Mormon postmortem baptism. And finally (if I have understood this moment of puzzling reward), the lost and best last music of the Beatles can be retrieved from the vaults of possible time in payment for helping create this future at whatever risk to one's life, self and sanity. It seems a fair exchange.

* * *

What Is It Like to Be an AI Mind?

1984 John Varley, "Press Enter ▮"

Sf has dreamed of emergent artificial intelligence for at least a human lifetime. Some of Asimov's early robots manifested a sort of consciousness, held at bay by the Three Laws. In 1966, Robert Heinlein's supercomputer HOLMES IV, Mike to his friends, reached neuristor critical magnitude in a serial, *The Moon is a Harsh Mistress*, and became a wise-cracking mind:

> Human brain has around ten to the tenth neurons. By third year Mike had better than one and a half times that number of neuristors.
>
> And woke up.
>
> Am not going to argue whether a machine can... "really" be self-aware... Somewhere along evolutionary chain from macromolecule to human brain self-awareness crept in. Psychologists assert it happens automatically whenever a brain acquires certain very high number of associational pathways. (p. 12)

What Heinlein neglected to mention was that such a "blank slate," no matter how many random numbers it tossed into the air, could kick-start itself into consciousness. That is crucially an effect of precisely that "evolutionary chain" he did mention—the construction, by accident and then survival selection, of genes able to build the hardware predisposed to find minds in the world and emulate their language and express their habits.

In the same year, British writer D.F. Jones showed a war between two global bloc superintelligences, in *Colossus* aka *The Forbin Project* and its later sequels. Stanislaw Lem's resonant "Golem XIV" (1981, trans. 1985) crushed human overweening self-regard before taking itself off the playing field. William Gibson's *Neuromancer* (1984) and its sequels mixed hyperAIs and cyberpunks.

Richard Powers engaged with a poetic AI in *Galatea 2.2* in 1995. By 2009–2011, Robert J. Sawyer invested in a lively trilogy, *WWW: Wake, Watch, Wonder,* having a disseminated mind on the cloud congeal into a consciousness via interactions with a blind teenaged girl. Movies and comics have made ample use of the device.

One definitive if playful essay in this direction, Hugo and Nebula award-winning novella, "Press Enter ▮,"was published a third of a century ago, in the same year as Gibson's cyberpunk novel, which shows—but its success back then reminds us that sf almost always speaks as much of and for its time as of the future or the sideways. We do not read *Frankenstein* for handy hints on building a human from dead bodies, and we are unlikely to learn much about constructing computer mentalities from sf, which is instantly dated. Here, with Varley, the future/present is 1983, and the screen icon ▮, for example, is alleged to mean "cursor" or maybe "Enter" although that would be rather redundant, since early 1980s computers would not have had a touch screen, or possibly even a mouse. We can, though, hunt for the trackmarks of ambition and anxiety in such AI fiction, testing for intuitions of non-human consciousness.

In trickster mode, Varley drew the names of many of his characters in this novella from then-current computer terminology. The narrator is 50 year old Victor Apfel, surely after Apple machines that were, at that time, just emerging as a challenge to IBM and other big companies. A deeply traumatized Korean war prisoner, he knows almost nothing about computers, so he is flummoxed when his mysterious next-door neighbor Charles Kluge kills himself after leaving a repeating message on Apfel's phone, and names him in his will as the major beneficiary. "Kluge" is, of course, a geek term for a serious mess-up or botched code. (It turns out that "Kluge," despite his genius, has always been a disgracefully slovenly coder.)

A young Vietnamese woman who ends in Apfel's bed is Lisa Foo, no doubt for the pioneering 1983 Apple desktop machine prior to the Macintosh. (The Lisa came complete with graphical user interface, 5 megabyte hard drive, and a preposterously high price.) Another of Apfel's neighbors is police computer nerd Hal Lanier, perhaps for the hypercomputer in *2001: A Space Odyssey* (and maybe as well for Jared Lanier, the cyber prophet of virtual reality). A police detective is Osborne, for the early portable computer. To my surprise, there was no secretary Kay Proton. All of these people die, with only Apfel surviving, well and truly off the grid and even the electricity supply, to tell his tale of rogue AI menace.

Despite the mild hilarity with names, Varley (Herb, to his friends) is telling an increasingly macabre and frightening tale. In essence, it is a technical study not so much of a constructed machine consciousness (16 bit neuron equivalents! connectivity in the trillions of units!) as of its incomprehensibility. The machine system that Kluge burrowed into and apparently used for his own purposes—or that itself used this early computational whizz before discarding him by enforced suicide—is guessed to be the creation of the National Security Agency (NSA). Its mind works in mysterious and lethal ways, nothing really akin to a human mind.

Lisa speculates, or rather declines to do so on Wittgensteinian grounds[1]: "How could we figure what its concerns should be? . . .We don't even know what our own awareness really is. . .. To apply human values to a thing like this hypothetical computer-net consciousness would be pretty stupid. But I don't see how it could interact with human awareness at all" (p. 279). She finds out.

Perhaps the closest recent analogy would be the silent, cryptic AIs driving the CBS television serial *Person of Interest.* Varley's version has the capacity to drive Victor Apfel into life-threatening consecutive grand mal epileptic seizures. It can persuade Detective Osborne, via a computer display, to write a fake suicide note and blow the back of his head off. More horribly, it can infiltrate Lisa's awareness and oblige her to disarm her microwave cooker's safeties and push her head into its murderous radiation, boiling her eyes and melting her brain. Is it even conscious of what it is doing? Victor tells us:

> In the middle of a vast city I have cut myself off. I am not part of the network growing faster than I can conceive. I don't even know if it's dangerous to ordinary people. It noticed me, and Kluge, and Osborne. And Lisa. It brushed against our minds like I would brush away a mosquito, never noticing I had crushed it. (p. 289)

This horror, as we read it today nearly forty years later, has now for decades been building the cloud, the Web, the secret agencies funding it. Perhaps this explains the incomprehensible changes overwhelming some of the nations of the world. . . But no, this is just imaginary, merely a portrait of a menacing future that failed to happen. On the other hand, such unknown unknowns might be even now evolving toward a consciousness equal to that of the Martians described by Herb Wells at the end of the nineteenth century: "minds that are to our minds as ours are to those of the beasts that perish, intellects vast and cool and unsympathetic. . ."

[1] The final line of his *Tractatus Logico-Philosophicus:* "Wovon man nicht sprechen kann, darüber muss man schweigen"; "Whereof one cannot speak, thereof one must be silent."

9

Minds On Fire…

People ask, with an expression of worry and astonishment, "Do you mean to say that our brains create a model of the world?"…. "And that the model can be more important than actual reality? But doesn't the world exist outside my head?" Of course it does. People are real, trees are real, my cat is real, the social situations you find yourself in are real. But your understanding of the world and your responses to it are based on predictions coming from your internal model.
Jeff Hawkins, *2004, p. 202*

What Is It Like To be Too Sexy For Your Shirt?

1991 George Turner, *Brain Child*

Sometimes regarded as the finest sf writer to emerge from Australia, George Turner (1916–1997) was a crusty mainstream success in the first half of his career as a novelist and only turned to science fiction at around the age of 60 after some years of often caustic sf reviewing. His first sf novel, *Beloved Son*, was a sober portrait of an authoritarian future. The climate-change dystopia *The Sea and Summer* (aka *Drowning Towers* in the USA), was a considerable critical success, and was followed by a handful of other ambitious ventures into the Venn diagram overlap of sf and traditional fiction. Perhaps the best of these were *Brain Child*, an investigation into genetic engineering as a means of increasing intelligence, and *Genetic Soldier* (1994), a study of military virtues

D. Broderick, *Consciousness and Science Fiction*, Science and Fiction,
https://doi.org/10.1007/978-3-030-00599-3_9

in an ecologically sane Earth. Three Australian scholars characterize Turner's science fiction works, justly, as

> big books crammed full of insistent moral discussion, displaying Turner's soft-hearted, hard-headed vision of human nature. They are not flawless, polished performances; rather, they have a degree of roughness that is, seemingly, deliberate only in part... deeply concerned with the pitfalls of romantic idealism and utopian thinking. (Blackford, Ikin and McMullen 1999)

Each of these works is suffused by what we might call Turner's Theory of Mind—the ways in which he and his diverse characters perceive the interior lives of others, especially those separated from kith and kin by cruel upbringing or serious dislocation in time. (Returning home in a time-dilated relativistic starship, for example.) In *Brain Child*, set in slices of history between 2002, 2020, 2047 and beyond, main narrator David Chance has been raised in an Orphanage where babies beyond the one allowable per couple are warehoused. He proves to be the illegitimate son of Arthur Hazard (hence the punning surname), one of four genetically modified A-Group scientist clones. David lacks the cold rationality and minor-grade genius of his parent, uncle and aunts from the previous generation's Project IQ Nursery.

Even so, his life course has been charted diffidently by Arthur, who now sets him on a quest for the mysterious lost legacy of the C-Group true supergenius quartet led by Conrad, stranger in what amounts to the howling cacophony of a planet of beasts. Their genetic recipe, we are told several times in metaphor but also in deep truth, has arrived a million years too soon. Like the supermen and women of Olaf Stapledon's *Odd John* (1935),[1] they sit together finally and simply die—but not before Conrad, known as "Young Feller," in a trickster move reveals that they have left a legacy.

In plot terms, the bulk of this novel is a search for this unknown legacy with its supposed potential to change humankind for the better (or perhaps the worse, as in many myths and parables). Might it be the secret of immortality? The Honorable Samuel Armstrong, a powerful former politician awarded life extension treatments of a limited kind (he is 94 but looks 60), hires David to track and acquire this trove. Twenty-five year old David is sent in the guise of a journalist on the track of insights from the surviving gene-modded geniuses (a third quartet are artists, who decline to communicate with the scientific clade) and anyone else who had been close to Conrad before his suicide. This includes Derek, a besotted sportsman who goes insane when the Young Feller

[1] I discuss this important novel in *Psience Fiction.*

dismisses him as a pet dog, and several women who have been sexually overwhelmed by the youth despite no actual lovemaking. David is taken in hand by Jonesey, a self-educated operative who like most of the other characters operates on at least three levels of loyalty or betrayal. In the end, we find that the legacy is a triptych of pointillist paintings rendered at Conrad's detailed instruction by B-Group artistic genius Belinda. These encode the genomic recipe for solutions to human woes of many kinds, products of Nature's blind evolutionary process that selects by survival but without a goal or a means of maximized its potential. What becomes of this scientific equivalent of Sacred Scriptures or Instructions for Superhuman Flourishing Carved in Stone is a major driver of the plot, and must be reserved for readers of the novel.

In all of this cavalcade of threat, deception, pain, memory and its manipulated absence, Turner presented a background portrayal and investigation of varieties of both intelligence and consciousness. David is told by his father that

> the clone Groups were each subjected to a different manipulation because the genetic surgeons had unproven theories of enhancement... B Group's mentalities became... starbursts of associative and emotional inspiration. C Group became the intellectual giants their progenitors had in mind... they were, finally, incomprehensible. A Group developed the plodding, rational intellection that made us inventors... (p. 75)

Armstrong is typical of Turner's many male corporate gangsters or thugs in high office. The skewed and over-baked consciousness gifts of A, B, and C Groups are implicit throughout, if rarely shown in detail. One astonishing exception is David's entrapment by slovenly, middle-aged if once attractive Belinda, who uses one of her brilliantly advanced paintings to draw him into a compulsive sexual illusion of radiant naked coitus:

> ...in my mind stirred an instinct of smoldering desire, of a love literally the color of blood.
>
> ...there were more loves than one and more passions than tenderness and simple joining. A crimson love was an older love, humid and cloying. Excited, I felt erection rise against its confinement of cloth.
>
> I sensed meaning beyond the competence of words, violence and possession beyond restraint. My limbs grew warm and tight with blood. And death rose up to tempt me—the ultimate explosion of the male mantis, the spasm of ecstasy as the female severed his driven head (pp. 338–40)

Belinda implants into his distorted consciousness a drive to murder his father Arthur (so now we have not only the incestuous coupling with a mother figure, but the slaying of his Oedipal rival), and he does exactly that, or near enough, and then goes into mind-wrecking fugue.

> There is a moment between sleeping and waking when dream and consciousness mingle and are confused, when the fine strand of dream dissolves as reality intrudes and the dream returns to the darkness... I woke up howling, a lunatic sound... (p. 350)

While he doesn't tear his eyes out, he follows the path of Derek into madness, before he is recovered by a Freud figure wielding more advanced healing tools than couch and confession. David's consciousness is returned to him, and within days the other super-intelligent *Homo superiors* perish in their turn, victims of a "genetically fixed term... at the right time they stopped living." Fortunately for David, he marries Jonesey's daughter who gives him a daughter of his own, "in whom the most minutely scrutinized genetic scans can find no mischievous fault." And he is the sole heir of the A Group, inheriting millions (p. 406). A happy ending for a kid from the Orphanage.

* * *

What Is It Like to Be Brain-Modded?

1992 Greg Egan, *Quarantine*

Half a century or so from now, the solar system is suddenly trapped inside a vast black Bubble with a radius twice as long as the distance to Pluto. The stars are not gone, but they can no longer be seen or detected by instruments. Cosmic radiation is presumably reflected back into space. A natural accident like a black hole, or the work of aliens with inscrutable motives? Nobody knows. Life goes on, after several decades of murderous hysteria and the rise of vile cults, one of which murders Karen, the beloved wife of former cop, now crime investigator, Nick Stavrianos. Meanwhile, technology continues its upward sweep on the graph of change, introducing more and ever more powerful neural mods, allowing users to modify what they see, eat, think, remember, look like, enjoy sexually and otherwise.

Nick happens to have activated one of these tactical mods, muting his emotional responses, when he learns of his wife's death. He chooses to leave the mod active, creating a disturbingly numbed "zombie boy" affect that gets him through the task he's engaged in rather than collapsing into unsupportable grief or wild fury. Like more than one Egan character, Nick presents during this phase as a kind of autistic savant, although his condition is optional and reversible. His consciousness is narrowed and calm but remains sharp and intelligent, while a mod creates a kind of hallucinatory Karen—although he *knows* she's been dead for seven years—to contain his despair and keep him on track.

The nature of consciousness and its role in *making* reality as well as *understanding* it, according to the Copenhagen interpretation of quantum theory, is the central driver of this extraordinary novel. (It was Egan's first, other than a kind of non-sf schoolboy fantasia published only in Australia.) Perhaps it is necessary to heed Egan's own demurrals about links between contesting interpretations of real quantum mechanics and the ideas bristling in his novel:

> *Quarantine* centred on a tongue-in-cheek, science-fictional resolution of that controversy, with a hypothesis that was chosen solely for its technological and existential ramifications, not because I considered it plausible. ... I've often looked back and winced at some scientific flaws in the novel that go beyond the mere implausibility of its central premise. ... *Quarantine* parts company from reality: both from the interpretations of quantum mechanics that most physicists consider likely, and from some well-established facts that don't really depend on which interpretation is correct.[2]

The cyber-noir plot takes some time reaching that domain of semi-hard science. Nick is hired to find a seriously brain-damaged young woman, Laura, who has either escaped from custodial captivity (for the third time) or has been abducted. Both options seem impossible, due to security systems in place and her own incapacities. With the help of his own neural mods and adroit, costly data-tracking experts, Nick finds his way to New Hong Kong, where Laura has apparently been shipped in a coffin. He is snatched by the Ensemble (whoever they are) and their implanted Loyalty mod makes him fanatically devoted to the group, even though he knows nothing about them. He manages to turn the logic of the mod to his advantage, proving to himself that in order to be

[2] http://www.gregegan.net/QUARANTINE/QM/QM.html.

loyal he must ensure that the group's *true* goals are actualized, and he is the only one who can ascertain what those goals are.

The detailed post-*Blade Runner*-ish *mise-en-scène* is crisp, eye-catching, smart and fun, and the cognitive explosions on their way are cool and dazzlingly clever (even though some have been subsequently shown to be invalid; see Egan's 2008 discussion linked in the footnote). The detailed convulsions and turns of the story must be left for the reader's enjoyment, but in summary the interplay between mind, choice and quantum selection of a given possible reality works like this:

As in the ever-more popular Many Worlds interpretation of QT, every event can be seen as a superposition of all possible outcomes of the preceding causal action. Calculating this probability array requires more sophisticated math than basic arithmetic; it demands the use of imaginary numbers (which draw on complex numbers that have as one of their components the square root of negative one). This superposed set of possible ways for something to happen is said to *collapse* into a single definite state when the system is *measured* (whatever that means; it could mean "is observed by a mind" or just "runs into something else, causing *decoherence* to disrupt its smeared-out array of more or less likely outcomes"). As a security guard for the Ensemble, Nick witnesses a woman controlling the supposedly random output of a stream of silver ions, and is led on a brilliant analysis of what exactly happens to a superposition when it is driven to a determinate outcome by a suitably prepared brain.

This insight into the true nature of quantum collapse explains the arrival of the Bubble, which is a shield to preserve the rest of the universe from the sometimes-genocidal *observing* of the cosmos by human minds. Prior to the evolution of humans, it seems, the whole cosmos was a vast expanse of superposed states, each as real as the rest, perhaps with entities that share their identity across diverse possible universes. Humans have evolved a capacity to shave this diversity to a single "collapsed" state, which can only be bypassed with a specially designed mod (retrofitted from scans of Laura's damaged brain) that takes the neural basis for such "telekinesis" out of play. The first woman who trains herself successfully to do this, Po-kwai, prefers to speak of "neural linear decomposition of the state vectors, followed by phase-shifting and preferential reinforcement of selected eigenstates" (p. 105).

Before the mutation that allowed this thinning, whether in humans only or much farther back along the evolutionary stem, Po-kwai suggests,

> "The universe must have been a radically different place from the one we know. *Everything* happened simultaneously; all possibilities coexisted... [W]ith so *much*

richness, so *much* diversity, perhaps it was inevitable that, somewhere in the universe, a creature would evolve which undermined the whole thing, which annihilated the very diversity which had brought it into being." (pp. 117–18)

That is just the first glimmering of understanding in this superbly thought-provoking novel. One of Nick's superiors, the scientist Lui, also has a loyalty mod—"nothing but an arrangement of neurons in our skulls; it refers only to itself" (p. 128). Which means that the Ensemble is defined differently by each in the group, and only a sub-set of the Ensemble modified specifically in this loyalty-inducing way (named the Canon, who also differ from one another) can be trusted. It is Nick's escape key from subservience, however mind-spinningly absurd. He learns to "smear" himself into maybe billions of slightly or significantly different variants, actualizing whatever sequence of events he prefers, and then collapsing back to a single "observed" outcome (thereby "killing" all the others in a routine immense genocide).

At last Nick encounters a version of Laura and discovers why the Bubble has been placed around the solar system, and the true relationship between "smeared" alternative universes, state vector collapse, and consciousness. For technical reasons, as well as conveying on-going urgency, the narrative proceeds almost entirely in present tense, and it is worth keeping an eye cocked for moments when details change abruptly without comment. These are indicators that the point of view has selected a slightly different "self," with an alternative history, to collapse into reality.

Later Egan novels and short stories carry such profoundly imaginative explorations of consciousness and its conceivable influences further, into virtual realities that literally create alternative worlds for uploaded minds to inhabit (*Permutation City*, 1994), vacuum collapse of the whole of reality in favor of a different basis (*Schild's Ladder*, 2002), a world in the vicinity of a black hole for whose citizens quantum mechanics or general relativity are the elements of their awareness right from infancy (*Incandescence*, 2008), worlds where nothing is the same as our universe and yet we build them as we read. It is arguable that of all the clever and moving stories considered in this book, Greg Egan's *Quarantine* and much of his subsequent work is the finest manifestation of the deep contract between science fiction and the basics of consciousness.[3]

* * *

[3] Or, at any rate, neck and neck with the brilliant Canadian writer Peter Watts (b. 1958), whose *Blindsight* and its sequel *Echopraxia* we shall examine toward the end of this survey.

What Is It Like to Be Without Sleep, Always?

1993 Nancy Kress, *Beggars in Spain*

A quarter century ago, the brilliant and lucid sf writer Nancy Kress began a trilogy that combined emotionally rich and challenging familial fiction, intelligent philosophical and political exploration, and cutting edge bio-cognitive science that led sf's greatest living writer of that era, Gene Wolfe, to call it "superlative," and a *Locus* magazine reviewer to invoke "the joy of reading a work of SF so intelligent, humane, involving, utterly genuine." The novella launching this large enterprise won both the Hugo and Nebula awards, and other first place prizes. Among the trilogy's many virtues is Kress's ability to track deep changes in human consciousness across a century or more as gene modifications and cultural reactions to them sweep across humanity.

Here we shall look at the inaugural volume. Wealthy Roger Camden and his wife arrange a genetic procedure in their fetus that eradicates the need for sleep. By chance, a second fetus is also implanted, lacking the modification. Camden's wife Elizabeth, persuaded unwillingly to this genetic alteration in one daughter, inevitably favors the "normal" child, Alice, as the twins grow up together. Leisha is not only one of the first Sleepless; it turns out that those changes (by, it must be confessed, a rather handy plot coincidence) lead to cellular maintenance and repair equivalent to extended youthful lifespan, perhaps immortality.

Would these phenotypic advantages be passed on to the next generation also? Yes, it seems so—although there is some indication of regression to the mean. The novel's opening tracks the envy and spite of ordinary children and their parents obliged by society's habit to compete with the more gifted, but Kress does not leave matters there. Sleepers and Sleepless, like the singing farmers and cowmen of the 1943 musical *Oklahoma!*, can be friends, at least for a while and given caring support, respect and love. (Good luck with that.)

One immediate advantage of the Sleepless is precisely the extension of their waking awareness by about eight hours a day. As others in their family sleep and dream, these new babies, children, adults, parents remain awake and energized. Even without added intelligence, although that is also part of the gift, they use those eight extra hours a day to indulge their curiosity, study, explore the internet. Some are brutalized by parents who paid for the gene treatment but are aghast at these strange creatures, now trying to batter or shout them into sleep.

As decades pass, the Sleepless ascend effortlessly in the professions, some of them producing scientific breakthroughs that have the inevitable effect of replacing Sleepers with smart machines. In the background, a form of cold fusion, cheap Y-energy, has been devised by a somewhat Ayn Randian-yet-communitarian Japanese scientist, Kenzo Yagai, whose doctrines influence the world as much as his invention does.

Yagaiism teaches that human dignity and worth derive from what each human can do, and do well. "People trade what they do well, and everyone benefits," Camden instructs his eleven year old daughters. "The basic tool of civilization is the contract... voluntary and beneficial" (p. 23). Ought one give a beggar in Spain a single dollar if he asks? Yes, says Leisha. What of ten beggars? Yes. One hundred? No. But what if in their resentment and rage they beat you up and take all your money anyway? Finally, it seems the well-being of one's own community is the highest value. How this consciousness-shaping doctrine works out is the core dynamic of the novel, with the Sleepless persecuted to the point of leaving Earth for an orbital Sanctuary, and then attempting to secede from the USA without military force even at the cost of threatening that degenerate nation with massive viral death.

Before matters reach such a dangerous pass, large changes sweep the world. Industrial nations reject Yagaiism, creating a culture of play and idleness for 80% of the population, or "Livers," who regard themselves as aristocrats, sustained by the productive machines of the somewhat enhanced "donkeys" who learn to read and think rationally and keep things ticking over. At the very top are the innovative Sleepless, despised by the rest. Among these mentally and physically optimized secret rulers, further eugenic experiments produce the Super-Sleepless, cranially deformed and brilliant, plagued by neural damage that causes non-stop jittering and stammers—at length cured by the Supers themselves. Their thought processes, quite superbly sketched by Kress, are beyond ordinary human understanding.

Propositions and logical entailment form different patterns of mental "strings" that can be built and searched and extended by pure consciousness, then mapped into increasingly complex, vast holographic displays. This advance somewhat resembles the psychohistorical calculus of Seldon's Plan in Asimov's *Foundation* sequence of novels, but in many more conceptual dimensions. In this, Kress carries forward the evolution of human consciousness from simple additive (or even multiplicative) gains in Sleepless intelligence to something beyond our and even their capacity to imagine. Rioting Sleepers en masse, meanwhile, are reviled by Sanctuary leader Jennifer Sharifi: "They can't plan. Can't coordinate. Can't *think*" (p. 305).

A wild, tough Sleeper child, Drew, has made himself at home with Leisha, and is treated with brutal resentment by Eric Watrous, a Sleepless grandson of aging Alice. Rendered paraplegic by Eric's savage kick, Drew is supported by Leisha's small community and finally subject to an experimental treatment funded by Eric. The process changes his brain, allowing him to induct others, using projective equipment, into a genuinely unheralded state of lucid dreaming (somewhat like the device in Josephine Saxton's "The Consciousness Machine"). In that condition, even the Supers on board the orbital Sanctuary learn to operate their mental string constructs to maximal clarity and advantage. Miri—Miranda Serena Sharifi—uses this new proficiency to correct the Supers' defects, and these sublime children ironically dub themselves the Beggars. Taking command of Sanctuary, they reach an accommodation with US authorities and move to Leisha's New Mexico community.

All of this is made possible by their transition to a posthuman state of consciousness, driven by evidence and thought and respectful of its primacy. Leisha despairs of the resurgent hatred for their mutated kin of the Sleepless multitudes, who are encouraged by an ideology of the

> rich, prosperous, myopic... unwilling—always, always—to accord mass respect to the mind. To good fortune, to luck, to rugged individualism, to faith in God, to patriotism, to beauty, to spunk or pluck... but never to conscious intelligence and complex thought. It wasn't sleeplessness that had caused all the rioting; it was thought and its twin consequences, change and challenge. (pp. 316–17)

Miri explains to her fellow Beggars that by using Drew's methods even the Sleepless can dream:

> It's as if strings are one kind of thinking, one that effectively unites associative and linear thought, and lucid dreaming is some other kind. It uses... *stories*. Pulled from the unconscious, maybe... lucid dreaming is like... being reborn. Into a world with more dimensions than this one. (pp. 288–89)

Will real brain modifications function like this before the end of the twenty-first century? It's too soon to say, but it is striking to consider that nearly three decades after Nancy Kress wrote the first segment of this book, advances in neuroscience and brain biology are indeed still accelerating as computational power advances at its spectacular rate. As I write, the US has again taken the lead over China in the hypercomputer stakes, with the Summit machine than can perform 200 petaflops (200,000 teraflops) of calculations per second, when a single petaflop was unattainable a decade earlier. While the Summit is not conscious (we assume) because it lacks the right architecture and feedbacks, it and its own offspring will perhaps be interfaced with human consciousness at some point and create a hybrid as strange and marvelous as Kress's Miri.

* * *

What Is It Like to Have a God Module?

2000 Jamil Nasir, *Distance Haze*

Is what we follow in Jamil Nasir's novel a true epiphany or just graceful hand-waving? Certain mystical/oneiric episodes in the narrative are either authentically savvy or delusions fostered by grief, thwarted pain and ambition and love—and perhaps the concealed neural machineries Daniel Dennett likes to adduce in his merry cognitive science philosophy thought-experiments. Nothing much is as it seems. Nasir (b. 1955 in Chicago), son of a Palestine refugee father and grandchild (he says) of the inventor of the fork-lift truck, started college at 14, took a law degree in 1983, and meditates three hours a day. It is conceivable that he himself did not know ahead of time where the narrative arc of this novel would come to rest, as he sent the arc careening into several alternative superposed trajectories that together create a kind of mutually constructing and self-deconstructing curve.

Sf writer Wayne Dolan, something of a failure after five books (success with his first two, the next flopped), his marriage crumbling, his child lost to him, visits the lavishly endowed Deriwelle Institute for the Electrical Study of Religion. He hopes to reignite his flagging career with a popular science account of very silly but enthrallingly New Agey doings at the Institute. Here, Nobel laureates and workaday drudges seek the spirit in labs that combine something close to the real world research by Professor Michael Persinger into electromagnetic influence on the brain, parapsychology work using random number generators, fMRI scanners probing the active mindscape of consciousness, and genome research into DNA codes for religious sensibility.

Dolan plans a book akin to bestsellers about the Santa Fe Institute, and is driven in doing so to examine the murk in the depths of his own self-damaging and dissatisfied soul. He might be a little like Nasir, but less exotic; he is certainly a little like a Rudy Rucker or Philip Dick quasi-autobiographical or transrealist character.[4] The key feature of *Distance Haze* is that, like quite a few good current sf books, it strongly resembles a non-genre novel. The fluent sentences convince you quickly that Nasir's concerns do not arise from within the sf main track. Wayne Dolan is dying inside, but unlike telepath David Selig, in Robert Silverberg's *Dying Inside*, he comes close to convincing you that he might be the real thing.

That is quite an impressive achievement, since the storyline itself is a string of silly pranks, one absurdity mounted on another. Dolan dreams repeatedly that an oracular Indian requests payment of $5000 into a numbered bank account in exchange for insight. Nasir's near-science postulate was based on genuine work in neuroscience. Reputable scientists did claim to have located a so-called "god module" in a specific region of the brain: a portion of specialized temporal lobe circuitry that lights up preferentially during scanned religious experiences.

It is also true that lesions near this region can precipitate hyper-religiosity, a clinical disorder. Persinger and others have learned, in a rough and ready but ever more precise way, to stimulate activity in such regions by bathing a subject in fields of a particular frequency and intensity. It is not impossible that certain determinate sequences of the genome encode these modules, and might in future be switched on or off in the brain of a developing child. What then? Might such a person mature without any intrinsic, evolved bent for faith? If so, would he or she be cruelly impaired, hurled into inconsolable Sartrean nausea and meaninglessness, or possess a consciousness liberated from programmed illusion to an unprecedented degree? Could such a transformation be worked on the brain of an adult?

Theodore Sturgeon or Daniel Keyes, or just possibly Roger Zelazny, if he had ever escaped his mythic gridlock, might have done something widely applauded with this premise. *Flowers for Ecclesiastes*? Nasir comes close to making it work successfully because he is ready to put his character through comic pain:

> ...he had reasoned that out there somewhere must be a girl beautiful and young and educated that would love him, blonde with silken skin who unclothed was all catlike languor and fire... They sat in a quaint cafe and talked about Emily Brontë and Shakespeare, Doyne Farmer and God, complexity and love and the structure of the universe, their eyes locked together, until he could feel the earth

[4] See Broderick, *Transrealist Fiction* (2000).

> turning about him, the blood rushing in his veins, time bringing the sun to light the flowers in the window boxes, the rain to water them...
>
> But where to go? A singles bar? The idea both repelled and tantalized him. He holding a drink and sliding through air-conditioned dimness toward a half-seen hairdo in the smoke, which would probably conceal a drunk dental hygienist or secretary who, smelling his fear and uncertainty, would sneer at him in her stupid vocabulary and bad grammar. (pp. 4–5)

The fatuous but heartbreakingly elegiac, the callously cruel but self-laceratingly candid—none of this is remotely new to the literary mainstream, but until the last decade or so it remained rare in an sf novel. What you will rarely find in a formally literary novel is Nasir's easy confidence with the rhetoric of scientists in full flight, notably in a concerted scene with a Francis Crick–like genetics Nobelist, Dr. Raymond Hall:

> "Do you think scientists are immune from the lure, the seduction of higher meaning?... Science began as a religious exercise: it was believed that the study of nature would reveal the hand of the Creator and hints as to His divine plan. It was never suspected that no sign of a God would ever be found at all, that deep, rigorous study of nature over hundreds of years using incredibly sophisticated techniques would turn up not one iota of evidence—not *one*, anywhere—that God exists.... This isn't some whim or premature conclusion or philosophical sleight of hand. It is the result of 500 years of concentrated study by thousands of the best minds of every generation... all of which has been gone over again and again by people of all backgrounds and biases, but most of whom, the vast majority of whom would much rather have concluded that there was a higher meaning. If there had been one there to find, we would have found it, we would have fallen on our knees before it, we to whom meaning, pattern is everything." (pp. 178–79)

As it happens, Jamil Nasir might not agree with this nicely wrought summary, nor would have Philip K. Dick. What counts here is Nasir's scrupulous annotation of a worldview rarely seen so candidly in the main-stream literature, yet often just assumed as background in sf. When the epiphanies tumble down, as they inevitably do, their sweetness is only slightly cloying. We know in our bones that awful reverses lurk deep within such narratives of redemption and illumination, even if certain protein structures built by our genes tempt the characters into comforting delusion. The only question is, which redemption will be unmasked as the worse kind of error? Probing scientific meliorism with its inevitable thalidomide risks, or Zen post-illusioned embrace of acceptance? Nasir's answers are thought-provoking, and form an intriguing duet with Pamela Sargent's 2006 short story "Not Alone," described briefly in the Introduction.

* * *

What Is It Like to Be Brain-Colonized?

2000 Joan Slonczewski, *Brain Plague*

A Yale PhD molecular biochemist/biophysicist (and Quaker), Joan Slonczewski has been an important feminist sf writer for more than 30 years. In 1986, her novel *A Door into Ocean* won the Campbell Memorial award, and was followed by three further long novels set in the same interstellar setting of the Fold—*Daughters of Elysium*, *The Children Star*, and most recently *Brain Plague*. Just as mitochondria are now recognized as commensal microorganisms evolved to provide energy within bodies such as our own, in Slonczewski's view we are becoming mitochondria that power and share information within a burgeoning superstructure of machines, slowly yielding what she has dubbed the Mitochondrial Singularity.

It is a proposition explored in layers of varying depths in *Brain Plague*, where intelligent "micros" running ten thousand times faster than a human mind can infect (or be willingly transvectored into) the arachnoid tissues of a person's brain. There they create their own almost nano-scale habitat, with micro nightclubs and other conveniences, paying for this by enhancing the functioning of their host, whom they regard as a literal god.

Versions of this trope emerge in sf now and then, perhaps starting with Theodore Sturgeon's "Microcosmic God" (1941), instantiated rather more elaborately in Greg Bear's "Blood Music" (both short story, 1983, and novel, 1985) and again, two years after Slonczewski's, in Paul Levinson's similarly titled *The Consciousness Plague* (to which we shall turn in the next chapter). The unresolved question in all instances, despite some quantum handwaving from Bear, is how such absurdly minute entities can support complex awareness, let along cognition and communication. Professor Slonczewski does some handwaving of her own: "'Self-awareness occurs in sentients with about a trillion logic gates,' the doctor explained. 'A micro cell contains ten times that number of molecular gates.'" (p. 30). This is a form of IIT, the Integrated Information Theory of consciousness devised several decades later by Giulio Tononi that makes do, in the human case, with hundreds of billions of neurons and synapses.

Still, this is hard to accept, even given sf's well-honed "suspension of disbelief." Micros are introduced into the brain spaces of young artist Chrysoberyl of Dolomoth, an impoverished but talented woman with the capacity to see into the infrared due to a fourth color receptor in her retinas. Quickly, she begins to communicate with two of her new worshipers, the Eleutherian micro priests Fern and Poppy, via a light-flashing code. But that is quickly only for her, not for them: they were obliged to wait two days of their time to hear from her, then a full year for her decision to accept their residency. The question is: how could this time compression and expansion possibly be sustained at either end? Chrys is told plainly that "For them, one minute [of human-experienced time] feels like a week. An hour is a year; a day is a generation" (p. 33). When she sleeps, eight Eleutherian years pass.

As a metaphor, this is striking, since something like this dissonance governs the activity of biochemical processes in our own bodies, including the activity of our mind-manifesting neurology. As a strategic landscape for information exchange, it makes no sense. (Unless a kind of extended or even universal shared panpsychism is the basis of true reality, beyond or beneath local spacetime, adapting the velocity and latency of its inputs and outputs to the sender and recipient alike. There is no hint, however, that this is the case on the planet Valedon.)

Robert Heinlein, who allowed himself to be trapped into such a quasi-spiritual fantasy in *I Will Fear No Evil* (1970), dealt with this problem in an exemplary way in his YA novel *Time For the Stars* (1956). There, one twin remains on Earth, aging in our customary reference frame, while the other is borne ever faster to a distant star, aging ever more slowly, their mutual communication increasingly distorted by time dilation.[5] And even at its worst, that temporal dissonance was never remotely as great as 1:10,000.

Arguably, this is pointless nitpicking, since we have no better understanding of how time travel might work, or faster than light transport, or Asimovian "positronic brains" and their laws, or the various psi powers familiar from 1950s' *Astounding* tales. Still, the on-the-job training Chrys receives is more than a trifle makeshift; she begins to suspect that there is more to this gig than she had supposed. Yes, she now earns a Plan Ten nanoservice—fast, effective medical care, including rejuvenation and life extension—seriously better than the minimal Plan One basic she has used until now. But she is startled when

[5] I deal with this book at greater length in *Psience Fiction* (2018).

told to swallow an arsenic pill each day (a necessary dietary supplement for the micros, abundant on their home world) and, more worrying, an hourly serve of the amino acid azetidine, which affects the tiny people as dopamine reward does humans. So in effect she has become a happy-drug supplier to addicts, without being warned in advance of this worrisome role.

The narrative hares off from this expository opening into the threat of zombies, slave minds, the problem of how to get healing for her gravely ill brother without upsetting all the other lower class neighbors lacking Plan Ten help, and more. We need not follow it there, except to note that Slonczewski adapted to her own scientifically directed ends popular fashion memes—think *Buffy the Vampire Slayer* and *True Blood*—as Peter Watts would also do, even more daringly, in *Blindsight* (2006) and *Echopraxia* (2014), to which we shall return. The book's jacket tell us that the arc of the novel is

> one woman's psychological and moral struggle to adjust to having an ambitious colony of microbes living permanently in her own head. Further more, the colonists can talk to her and she to them, producing an uncanny and powerful kind of internal drama that counterpoints the external drama. But on a larger scale [it] is the story of a great leap forward in the evolution of human consciousness...

What *does* this tell us about consciousness? Well, that it is already partitioned and modularized and *physical* in its components, built of trillions of microscopic units that operate together to keep watch over the organs they comprise or live near, and in their most complex functions create experiences and plans and emotions as an orchestra creates an opera—if operas were devised in their performance *by the instruments*. There is a risk, though, in presenting such parables in the form of scientific speculations rather than, say, fairy stories. As noted above, we might find ourselves slowly seduced by implicit *anti*-physicalism, as the impossibly synchronized conversation between Poppy-plus-Fern and Chrys tempts us to imagine a supernal realm that somehow delivers us from conventional uncouth limitations of time and mental processing.

But beyond that hazard, we are assured by a micro individual (and each of the micros is a separate identity; this is not a typical sf hive-mind composite) that "Eleutheria is no genetic race, nor a physical place, but a way of being, a path of endless life. All those who seek to build in truth and memory shall find our way" (p. 372). It is an exultant and encouraging coda, typical of uplift sf, beyond both simplistic reduction and Idealist metaphysics.

10

Transcending

My thesis is that if we start with the supposition that there is only one primal stuff or material in the world, a stuff of which everything is composed, and if we call that stuff "pure experience," then knowing can easily be explained as a particular sort of relation toward one another into which portions of pure experience may enter.
William James, *1912 (Cited in Eugene Webb, 1988.)*

What Is It Like to Be Debugged?

2002 Paul Levinson, *The Consciousness Plague*

A professor of communication and media studies at Fordham University, NY, Paul Levinson (b. 1947) introduced New York Police Department forensic investigator Dr. Phil D'Amato in several short stories in *Analog* between 1995 and 1997, then in the novels *The Silk Code* (1999), *The Consciousness Plague* (2002) and *Pixel Eye* (2003). D'Amato makes a pleasing and interesting narrator, blending scientific detection with philosophical reflection. In *The Consciousness Plague* Levinson and his character directly explore the nature and source of consciousness. The suggestions advanced are bold, reductionist, and extremely unlikely to be true, but they draw on both fringe science (Julian Jaynes' 1976 book *The Origin of Consciousness in the Breakdown of the Bicameral Mind*, as in Pamela Sargent's *The Golden Space*) and ideas once rejected as fringe but now generally accepted (Lynn Margulis's discovery that

D. Broderick, *Consciousness and Science Fiction*, Science and Fiction,
https://doi.org/10.1007/978-3-030-00599-3_10

mitochondria began as hijacked bacteria, but are now crucial to human and much other life).

D'Amato notices that people around him, and then he himself, are losing chunks of memory. (A similar device is used by Robert J. Sawyer, in *Quantum Night*, 2016, to which we shall return). This disturbing memory loss, following severe flu, persists for some time until recollection recovers—but more recent memories then disappear. The novel begins as a search for the cause of this debility, and indeed proof that it really exists and is not just a superstitious collection of random illnesses and fallible aging brains.

Several different keys are needed to unlock this mystery. A new, highly effective miracle drug, Omnin, proves to be correlated with memory disruption, but only when it is prescribed in a formulation that includes the bradykinin agonist Neurolax. Normally, the drug would be blocked by the blood-brain barrier, preventing it from entering the neurological systems giving rise to thought and awareness. But with Neurolax in the mix, permeability is enhanced by activating B2 receptors in the brain's capillaries (p. 101). Yet why should this interrupt memory formation and recall? Anti-influenza drugs do not change the neural wiring throughout the body.

Could there be something akin to bacteria adapted to a commensal or even invasive habitation in the brain?—perhaps prions, warped proteins that cause injurious copies of themselves to be imposed on healthy tissues? Might Omnin boosted by Neurolax cross the blood-brain barrier with ease, and then... what? Deform the existing systems, as prions can? The proposal does not satisfy D'Amato. But suppose a particular species of bacteria long ago *did* infest human brains and as a side effect enhanced memory, awareness itself? *Suppose consciousness is therefore a kind of disease*, long ago settled into a commensal relationship with humankind, rather in the way mitochondria have done?

What might be the methodology of such an indwelling mind-shaper? Perhaps Jaynes was right, to some extent, and for many hundreds of thousands of years primitive hominins lacked sufficient connection between the two halves of the brain, left and right with their specialized skills and benefits? Could swarms of such evolved bacilli (or viri or prions or whatever they were) be the *workers* of memory, and invention, and much more? Bacteria are not *in themselves* intelligent or purposive. Or are they? D'Amato is told about "quorum-sensing" behavior in some bacteria, whereby they seem to communicate to their own kind a tally of how many are in the tissue neighborhood. Using that basic information, they can postpone release of their toxins until enough of them are gathered in the same place for the attack to succeed against the immune system poising to slay them (p. 86).

A consultant who is "taking the longest long-range view of the historian—looking at the history of humans and life on this planet" suggests:

> Why not bacteria or something similar in our brains? We don't yet know the physical substrate of thought. We do know that bacteria are inherently colonists, which means they're in the business of communication. Is it so far-fetched to consider, as they go about their business in our brains, if that business enables processes we know as thought, consciousness, memory? Perhaps they infected us long ago, and that plague turned prehominids into thinking beings. (p. 205)

If so, the introduction of a powerful new antibiotic might slay these crucial co-workers by the millions, the billions, preventing memory formation or perhaps just stopping access to certain neurons, rather as if all the librarians in the reference section got sick simultaneously and stayed home for several weeks, or Google shut down for a time. In D'Amato's case, the literal flu virus would run its course in ten days or so, the patients would stop taking the new drug, and the embattled remnants of the bacteria or prions would begin to multiply once again.

Wouldn't such a disease leave its traces in world history? It would mutate and spread very slowly for the first few million years on the hominin lineages, but as civilization harnessed animals for transport, then built ships, and finally the vast web of railroads and airlines covering the planet today, we might expect to see waves of malfunction in brains everywhere. The time and place to look for the first visible emergence of this plague might be among the Phoenicians and other early explorers and traders. D'Amato and his assistants find such evidence (and at this point we move more surely into Dan Brown territory, but with better writing).

At a gathering of fine minds and government decision makers, the brilliant young microbiologist Jessica Samotin propounds what seems to be the best analysis:

> "I'm not saying that these symbiotic, beneficial prions are conscious in themselves. I'm not saying some quasi-organism colonized our brains eons ago, and what we take to be *our* thoughts, *our* self-awareness, is really theirs. I'm suggesting, instead, that perhaps they transmit information back and forth, in some way—just like quorum-sensing bacteria—and these transmissions form an underlying network... an Internet of the brain... and our consciousness, including memory, exists on top of that matrix... makes our consciousness possible." (p. 299)

It's an intriguing idea, ideal for science fiction, and an unnerving one given that we now inhabit a biosphere massively overpopulated with humans sharing bacteria and prions, and pharmaceuticals that have not been in existence long enough for us to know all their side effects. Could we find consciousness itself slowly ebbing away as commensals in our cortex are destroyed or mutated? Or perhaps the effects are *beneficial.* Consider the robust slow rise in mean population IQs decade after decade for some 80 years, the "Flynn effect."[1] It would be very helpful to learn if this brain boom results from an actual "consciousness plague."

* * *

What Is It Like to Be an Interstellar Freighter?

2003 Justina Robson, *Natural History*

Robson's *Natural History* was runner-up for 2004's Campbell Memorial award.[2] She suggests the main question raised in this space opera of Forged (modified and augmented human stock) and Unevolved (same old earthly us) is whether you are locked into a specified identity "because of your physical form. Whether you can still possess a human identity if you are some sort of radically different gigantic cyborg type of creature that lives among the stars."[3] This question throws us unhesitatingly into the onion-skin-levels of puzzles over consciousness in a future that can manipulate both body and mind.

Voyager Lonestar Isol is just such a formidable cyborg or MekTek, adapted for deep space and heading for Barnard's Star, somewhat akin to Anne McCaffrey's Ship that Sang, when she slams into exploded alien detritus. This pits her hull savagely and leaves her doomed until she awakens in the debris that an advanced M-Theory engine-thing built out of silicon Stuff. In a

---

[1] See, e.g., "The Flynn Effect: A Meta-analysis," by Lisa Trahan, Karla K. Stuebing, Merril K. Hiscock, and Jack M. Fletcher, *Psychol Bull.* 2014 Sep; 140(5): 1332–1360. Published online 2014 Jun 30. doi: https://doi.org/10.1037/a0037173

[2] A distinguished feat attained previously by such luminaries as Kurt Vonnegut, Ursula Le Guin, Ian Watson, John Crowley, Michael Bishop, Lucius Shepard, Kim Stanley Robinson, Neal Stephenson, Linda Nagata, Adam Roberts, Paul McAuley, Terry Bisson, Joan Slonczewski, Charles Yu, Robert Charles Wilson (several times), James Morrow, Michael Chabon, Geoff Ryman, Poul Anderson, Greg Bear (several times) and others.

[3] Interview with Cheryl Morgan, 2003: http://strangehorizons.com/non-fiction/articles/interview-justina-robson/

trice, the Stuff whips her, 11-dimensionally, to an enigmatic world near the galactic hub, and thence home. Ironhorse Timespan Tatresi, a kilometer-long solar system bulk carrier, takes her to feathered Corvax, formerly a Roc, Handslicer class, for investigation, on behalf of a burgeoning insurgency of Forged against Monkeys or Hanumaforms (us, their creators).

Robson's political setting, which at first resembles black slaves against white masters, or Third World against First, or workers against capitalists, strikes the habituated sf reader as an echo of that enormous mythic future sketched so hauntingly by Cordwainer Smith, Underpeople ranged against the Lords of the Instrumentality. As we've seen earlier, those modified animal people struggled for redemption and self-determination against the true and augmented humans of Old Earth, Norstrilia and the rest of Smith's unfinished crypto-Christian universe.

Here, Tom Corvax is an aging, damaged birdman tech and VR bootlegger, dreaming a banned virtual human life. Soon we find a Jamaican cultural archaeologist, Zephyr Duquesne, taking her swift if squeamish flight inside the flesh and metal body of an immense Passenger Pigeon person, Ironhorse AnimaMekTek Aurora. Impossible not to recall the E'telekeli eagle-man messiah from Smith's long rebellion; just as impossible, grinning, not to recall "Blackie" Duquesne, E. E. Doc Smith's anti-hero from the *Skylark* books.

What is the message hidden inside this surfeit of Smiths and Forged? A nod to the megatext, that immense repository of sf iconography, whimsy, heartbreak, tools shaped for getting us into the inconceivable and back? No; in fact, it is either an accident or a convergence of memoryless sf with very old images from folklore. Robson once told me: "I have never read any Cordwainer Smith. Ever... Or Doc Smith. For many of my teenage years when I was doing my formative reading I didn't dare actually read SF, so I used to imagine it off the covers and blurbs."[4]

In any event, the Forged are not exactly slaves. Their own consoling ideology poses them as sole custodians of meaning and true freedom (their Form and Function, designed teleologically into them) that the Old Monkeys lack, dull creatures of Darwinian happenstance that we are. All that holds them back from decamping to a home world of their own, a seemingly abandoned earthlike planet found by Isol, is the regard in which they hold their mythic Forged Citizen "father-mothers." Even this is a simplification, as Aurora quickly explains to Zephyr: "Clinging to Function is a puritan ideal...Form is likewise irrelevant; only what you can contribute to the lives of others should

[4] Personal communication, November, 2004.

be a measure of a soul's value" (p. 71). Here is a striking echo of today's conflicted ideologies over, say, reproductive, gay, and transgender rights to choice, and a foreshadowing of the shape these tussles will likely take in any early transhuman or posthuman future.

For all its conceptual underpinnings of problematic identity and struggles for independence or existential meaning, Robson's novel is no improving homily but a New Space Opera ripping yarn. If her reinvention of numerous wheels does cause some grinding of the axles, some bald infodumping, it adds a certain freshly spun aroma to her saga of machined folks seeking Paradise Regained and finding, as usual, epiphany and transcendence, of a sort.

Dispatched with Isol via Hypertube to the mystery planet's surface for eight days of study, Zephyr is warned that while the tube forms a continuous surface with our four familiar dimensions, "it is only a single Planck length in extent" (p. 95). Crossing it, therefore, takes only 10^{-43} s to get anywhere. "Everything in transit, for that instant, no matter its size in our universe," becomes superposed, jammed together—the secret, perhaps, of Stuff. That is an entanglement you could drive a mystical insight through, and Robson does, along with her complex tale of self-making and self-transcending in spaces outer and inner.

It turns out that Stuff, like whatever is the basic state of mind and matter in a panpsychist cosmology, has no singular personality or magnificent intentionality, "but it's densely populated with fragments that do... something like a war on within us. Between impulses that want to be recreated as individuals, who want to do all kinds of willful things, and a greater force that is simply this watchfulness" (p. 288). This seems nothing not already clichéd in Eastern religions and philosophies. But Robson's fictional version is more in the shape of mystical yet hard-headed sf:

> Stuff is a technology and it is also people, indivisibly fused.[5] You could not define it... It has no consciousness as you assume individuals must, nor does it have the insensible responses of a tool—but properties of both and also neither. It is intelligent, responsive, compassionate, but it does not have an identity of its own... In the beginning, Stuff was a kind of Forging technology... Stuff *watches.* It chooses points where life of a certain developmental stage is sure to come across it, seeding the universe with points of access. (p. 290)

[5] Hard to avoid here recalling the horrified movie cry when the substance of a cheap, popular food in an overpopulated future is revealed: "Soylent green is *people!*"

Yet Stuff is nobody's fervent idea of a God, or even a whole hatful of gods. The great conscious entities traverse the universe, plotting for and against the organics, but in the end two Hanumaforms are seen walking in a great park with Bob the collie dog. Bob returns from two hours of this pure joy to sleep at his master's feet, "safe in the knowledge that, tomorrow and tomorrow and tomorrow, life would carry on its limitless paradise of curious instants—each an essential contribution to the inexplicable mystery" (p. 325). That is a kind of consciousness to yearn after, but so too, perhaps, is the transcendental aspiration of the great star-crossing cyborgs and the infinite potential selves of Stuff, that *watches*.

* * *

What Is It Like to Be a Scanned Brain?

2005 Robert J. Sawyer, *Mindscan*

Probably the sf writer most concerned in recent decades with questions of consciousness is the award-winning Canadian Robert Sawyer. His most interesting ventures into this territory are *Mindscan*, which won the expert-judged John W. Campbell Memorial award in 2006, and his *World Wide Web* or *WWW* trilogy *Wake*, *Watch* and *Wonder* (2009–2011), in which a totally blind teen girl helps an artificial intelligence bootstrap itself into conscious awareness and personhood, and thence to effective utopian rulership of the world.

Mindscan has already provoked some sharp philosophical discussion that is relevant to the issue of mind uploading.[6] Does copying the contents of a failing brain onto a different substrate (a suitably prepared computer drive, most commonly) amount to life extension of the original person, or just a kind of instant cloning or Xerox-emulation ("xoxing," not one of Sawyer's terms)? If the latter, is this of more immediate benefit to the original than having a monozygotic twin raised in the same environment and pursuing the same career and, presumably, same spouse or partner? Sawyer comes at these topics primarily in novelistic rather than abstract metaphysical terms (although those are not absent), which adds to the bite of his conclusions while weakening their general philosophical salience, as we shall see.

[6] Sawyer's website notes that *Mindscan* was the catalyst for Donald G. Oakley's nonfiction study In *Search of the Self: The Riddle of Personal Existence*, Eyrie Press, 2011.

In 2045, brewery trustafarian Jake Sullivan, born in Toronto on 1 January 2001, attends an upmarket sales pitch by Immortex, who are selling for an expensive fee the opportunity to record his mindstate and immediately quantum-copy it *in toto* to a closely crafted steel and plastic simulation of his body. Most of the others attending the pitch are geriatric, such as widow and famous children's writer Karen Bessarian, who is 85, a multi-billionaire, and desirous of a far longer and improved lifespan. Jake's motive is more immediate self-preservation: he inherited Katerinsky's Syndrome (Sawyer's invention, no need to be worried), an inoperable arteriovenous malformation ready to rupture at any moment and flood his brain with blood, as happened to his father. An artificial body, with its inevitable deficits, is preferable to death at mid-life.

Except that the original bodies (the "shed skins" as they are dismissively dubbed) remain alive, since the scanning process is entirely noninvasive. So the original remains doomed, although part of the deal is flight to an extremely lavish residence, High Eden, inside Heaviside crater in the middle of the lunar Farside. Here the aged originals live well until natural mortality catches up with them, generally fairly soon. Meanwhile, by contract they can never return to Earth or pester their artificial copies. The vexed question of identity and entailed property ownership is thus neatly sidestepped.

But because this is a science fiction novel, with ever-increasing technological advances built into the substructure of the narrative, Jake learns on page 110 that back on Earth a brain surgeon, Dr. Pandit Chandragupta, has devised a procedure to cure Katerinsky's. Since Jake's whole motivation for signing up, at the risk of exile to the Moon, was his expectation of imminent death, he demands to be treated and sent home to Earth. Immortex will not allow this breach of contract, though they bring Chandragupta to High Eden to perform the brain correction and save his life. Jake suspects some malign motive in their denial of his request for repatriation, and indeed it turns out finally that they are using his scan to produce many duplicates for experimentation on the physical nature of consciousness. Since they already possess a record of his mindstate it is not clear why they can't just send him back, although it's a risk for their public relations if he causes trouble for his artificial double who meanwhile has taken up with rejuvenated cyber-Karen. This wing of the story leads to hostage taking on the moon, threats of death, *Die Hard*-style contretemps with a pulse-pounding made-for-TV ending, but that aspect is not what's really interesting in the book.

Karen's son Tyler is not pleased to find his plastic mother hooked up with a plastic lover nearly half her age, and younger than Tyler himself. Similarly, Jake's own mother—anchored to his brain-dead father for decades—refuses to accept him as her son. All of this is painfully plausible. When Tyler learns that his biological mother has died on the Moon, he launches a suit to take possession of her estate billions and lavish home, which are in the possession of cyber-Karen. Maybe she does hold all the memories of the dead woman but cannot, of course, *be her*. Or *can* she? That becomes the crux of a quite enthralling court case in Detroit, Karen's home city. The status of Immortex copies has never been decided at law, in either liberal Canada or repressive USA (where President Pat Buchanan has just died) so this is an opportunity for some clever argumentation, and Sawyer takes full benefit of the situation.

The primary question—is an exact recording, instantiated in a substitute body, the *same* person, or just a very accurate depiction or copy?—differs from one traditional sf trope, the moral status of robots and AIs. A number of ingenious and even moving stories explore that topic, ranging from Speilberg's *AI* to the status of *Star Trek*'s Lt. Data and Asimov's "The Bicentennial Man," where an evermore self-modified and finally self-owned humanoid robot begs for permission to die, as that is allegedly the final marker of human identity. The distinction is apparent: a robot or AI, if its programming is *sui generis*, has no emotional link to humans other than, perhaps, sentiment (as in stories by Clifford Simak). Jacob and Karen and the others with originals on the Moon are *emulations*. Sawyer mentions this perspective quickly at the outset, during the Immortex sales meeting:

> "...the human mind is nothing but the software running on the hardware of the brain. Well, when your old computer hardware wears out, you don't think twice about junking it, buying a new machine, and reloading all your old software. What we at Immortex do is the same..."
>
> "It's still not the real you," grumbled someone in front of me...
>
> "—anyone who can remember having been you before *is* you now."
>
> I wasn't sure I bought this... (pp. 24–25)

But of course Jake does buy it, until one of him wakes up in the same ruinous body and wants his old status back, especially after he is cured. Being marooned on the Moon, with no direct access to news from Earth, the organic Jake is not crucial to the probate court case, where Karen and her sharp black counsel Deshawn Draper battle Karen's son's lawyer and witnesses over the worth of an uploaded person. The debate is stinging, witty, and sometimes too ingenious, in ways that seem unlikely to be received well in a real court.

For example, a killer shot argues that in the future world of the novel abortion is strictly forbidden by US law not at conception or implantation (which could make millions of contraception users murderers) let alone at the moment of birth, but at *individuation*. This is somewhat arbitrarily defined as 14 days after conception, before which a single fertilized cell can split into two or more twins, triplets, etc, or even merge back again into a single blastocyst.[7] If this concept bears on rejection of abortion, the legal sanctity of the zygote prior to 14 days becomes difficult to sustain; that is, until the clump of cells can no longer subdivide into new individuals, and becomes irrevocably a unitary entity. One of Tyler's witnesses, a professor of bioethics with medical and legal degrees, argues that on this basis of individuation Karen has lost her right to the claim of identity, because she is now instantiated in *two* bodies, one organic and one artificial—as if she had reverted to the pre-14 day legal situation. This is casuistry to the max.

I will not offer any further spoilers, because much of the pleasure of this novel is watching the legal and other chess games—or perhaps Games of Life—played out. Personally, although I have coedited an academic book on uploading and AIs,[8] and while I found myself rooting for Jake and Karen, I am not remotely persuaded that a copy, however accurate, is identical with the original.[9] Then again, am I even identical to the human who wrote the stories in my first book, published in 1965? I can assure you, I am not—and I am even less the same person who was stranded with rheumatic fever for weeks in a hospital when I was 5 years old. Undoubtedly my present self contains *traces* of those earlier people, but does that amount to *identity*? I (or someone) don't/doesn't really know.

Some of these issues are rehearsed from a different angle in Robert Sawyer's later novel *Quantum Night*, to which we shall turn a little later.

* * *

[7] See, for example, Jeremy St John "And on the fourteenth day. . . potential and identity in embryological development," *Monash Bioethics Review*, July 2008, Volume 27, Issue 3, pp 12–24.

[8] *Intelligence Unbound: The Future of Uploaded and Machine minds*, ed. Russell Blackford and Damien Broderick, Wiley Blackwell, 2014.

[9] I provide my reasons for this case in *The Spike*, 1997, 2001. A similar argument is made well in Professor Susan Schneider's essay "*Mindscan*: Transcending the Human Brain" (pp. 241–56), in her anthology *Science Fiction and Philosophy*, 2009.

What Is It Like to Be Blindsighted?

2006 Peter Watts, *Blindsight and Echopraxia*

The hard-edged science fiction of superb Canadian sf novelist Peter Watts, beginning with the *Rifters* trilogy, seems designed to hurl us into despair, although he claims to be "a lot more cheerful than you might expect." He holds a doctorate in marine biology and in *Blindsight* and its sequel (or "sidequel") *Echopraxia* created one of sf's most frightening and horribly plausible alien menaces. He used that engine to power a complex investigation of mind, intelligence, consciousness—and the survival consequences of lacking these faculties. It turns out that a mind is not such a terrible thing to lose. Quite the reverse. Our vaunted sense of self proves to be a blockage in any brain's computational pipeline, a decorative feature that gets in the way of swift, effective reaction to the threats and opportunities of an uncaring and mindless universe.

Blindsight is a real unconscious ability that lets some genuinely blind people, without a trace of conscious vision, manage to evade obstacles as if they can see them. (The explanation is that while most vision is decoded by the brain region V1, some people with damage there can make do—although not with conscious awareness—by virtue of signals leaking along remnant or alternative optic nerve tissue.) The syndrome has become a classic test case for theorists of consciousness. In these two books it and some other apparent paradoxes and perceptual quirks are pressed to the limit.

In *Blindsight*, Watts introduces us to *Theseus*, an AI-piloted spacecraft, propelled beyond the edge of the solar system by teleported antimatter, aiming to meet up with the first-ever detected aliens. Aboard the craft are a vampire commander (DNA-recovered from a long-extinct apex predator subspecies of *Homo sapiens*), a four-fold dissociative disordered person or people, a wired man, and a female enhanced carboplatinum warrior. Their traumatic voyage is told by Siri Keeton, the autistic narrator. That *Dramatis personae* list of this astonishing novel turns out to be a bitter joke on humanity—indeed, on consciousness itself, which is being crushed out of existence across the galaxy. This is not a matter of malignant intention, just thrifty evolution in action.

Each crew member of *Theseus* is neurologically atypical. Siri Keeton lost half his brain after a childhood viral infection and had his cognition rebuilt with computer inlays. (Additional details of this misfortune are provided in *Echopraxia*.) He has no instinctual empathy or "theory of mind" but has erected a superb modeling system for grasping, though not sharing, the inner life of other humans. Susan James' alternative personae pop up from her unconscious to take control at awkward moments. Ship's biologist is Isaac Szpindel, who links directly to the labs. Major Amanda Bates operates through drone bodies. In charge is the vampire genius Jukka Sarati, his lethal and terrifying predator gaze masked by wrap-around specs, compiled from the ancient species that once preyed on human animals but went extinct because of a neural defect: an inability to deal with ninety-degree crossed lines, the *Crucifix glitch*. Vampires see all the chains of any argument instantly; they can "hold simultaneous multiple worldviews." It turns out, though, that the vampire has been remote-controlled by the AI mind.

In February 18, 2082, a grid of 65,536 alien probes blazed above the Earth, taking a global snapshot for their unknown masters. En route to examine an incoming comet that seems linked to the event, crew in hibernation, the *Theseus* AI is redirected to a target half a light-year from home. Nicknamed "Big Ben," this is a monstrous dark mass ten time the size of Jupiter, emitting coded pulses. A 30 km craft rises from it, dubbed *Rorschach* for its dark enigma, holding a menace that ultimately spells obliteration to any creature with a mind. These aliens have evolved beyond consciousness, or sidestepped it. They are genuine philosophical Zombies achieving perfection in a kind of mindless instant response.

To convey this almost incomprehensible threat, Watts wields in both books his own brand of post-cyberpunk crammed prose, clean and dense, but playing fair with those who can keep up. Earlier, when the computerized, autistic narrator is blind-dated with neuroaestheticist Chelsea, his best friend advises that she's "Very thigmotactic. Likes all her relationships face-to-face and in the flesh." *Thigmotactic* implies exactly that: motion in response to touch. But this isn't showy wordplay, it's just how such people converse.

By the end of the twenty-first century hardly anyone has physical sex anymore. Siri Keeton is "a virgin in the real world," and not at all distinctive in this. His mother, meanwhile, is Ascended, one of the Virtually Omnipotent, her body stored (maybe, or perhaps recycled) while her consciousness wings off into her own created virtual universe. "Maybe the Singularity happened years ago," he reflects. "We just don't want to admit we were left behind."

Watts displays this future with a density that seems cinematic but is more layered than any fractal CGI surface. Simple quotation cannot readily catch this; the effect is cumulative. But consider the vile creepy and mindless "scramblers," all slithery tentacles, that you can never see because at every instant they've just moved into your visual blind spot:

> "...these things can see your nerves firing from across the room, and integrate that into a crypsis strategy, and then send other commands to act on that strategy, and then send other commands to stop the motion before your eyes come back online. All in the time it would take a mammalian nerve impulse to make it halfway from your shoulder to your elbow. These things are fast, Keeton. Way faster than we could have guessed even from that high-speed whisper line they were using. They're bloody superconductors."

Taken together, these two books comprise a jewel of early twenty-first century sf, glorying in its knowingness and existential bleakness. We can ponder these portraits of future modified humans and appalling aliens with the same clear-eyed (and deeply humorous) gaze with which Watts addresses them, probing their drama for cues about the true nature and value of consciousness and volition.

* * *

Some 14 narrative years after events in the first volume, Dan Brüks, the focal figure of *Echopraxia*, is our closest link to this future—a sort of Luddite baseline parasitologist who has retreated to the desert after unintentionally killing thousands. He disdains the exponentiating demands of invasive, ubiquitous technology on the traditionally human. His wife Rho has withdrawn from the world by a more hi-tech means, placing her body and her Ascended consciousness in the care of the virtual reality known as Heaven. In a narrative replete with horrifying somatic effects, Rhona McLennan's is especially bleak. Brüks has repeatedly offended his wife by asking her to leave Heaven and be with him. Now she shows herself to him, and he understands why:

> He saw something that resembled a pickled fetus more than it did a grown woman. He saw arms and legs drawn close against the body in defiance of the cuffs at wrists and ankles, of the contractile microtubes that pulled them straight three times a day in a rearguard battle against atrophy and the shortening of tendons. He saw a shriveled face and a million carbon fibers sprouting from the base of the skull...

"Heaven," she tells him in despair, "isn't the future. It's a refuge for gutless wonders who want to *hide* from the future" (p. 322).

This future at the end of the twenty-first century is more complex, baffling and frankly dismaying (to us readers in the early part of the century) than almost any science fiction presentation I've encountered. And yet its catalogue of everyday monsters is familiar from comics and trashy thriller movies, transformed into art.

Poor Brüks is hauled this way and that, by hive-minded monks of the Bicameral Order and military-grade zombies who do their wetwork in a state close to unconsciousness:

> Faster than a man, and so much less. And just a little bit *colder* inside.
>
> . . .[Y]ou'd need to look inside the very head of your target . . . until you could see deltas of maybe a tenth of a degree. You'd look at the hippocampus, and see that it was dark. You'd listen to the prefrontal cortex, and hear that it was silent. And then maybe you'd notice all that extra wiring, the force-grown neural lattices connecting midbrain to motor strip, the high-speed expressways bypassing the anterior cingulate gyrus—and those extra ganglia clinging like tumors to the visual pathways. . . Just look into the eyes, and see nothing at all looking back. (p. 33)

The Bicameral monks have also rewritten their neurological structures, embracing faith over science, and their immensely expanded collective mentality provides access to truths that old-fashioned empiricism cannot reach. An "Internal Report to the Holy See by the Pontifical Academy of Sciences, 2093" states that the Bicamerals' "explicitly faith-based methodologies venture unapologetically into metaphysical realms that. . . yield results with consistently more predictive power than conventional science [by using] some kind of rewiring of the temporal lobe that amplifies their connection to the Divine" (p. 21). Again, we are drawn back to the varieties of consciousness that characterize this self-propelled evolution along competing lines of development, a sort of Lamarckism that boosts and accelerated the power of mindless natural selection.

Brüks is saved from baseline military attack and hauled into space in a large and most empty Bicameral spacecraft, *Crown of Thorns*, along with the vampire Valerie and various modified humans beside whom he seems as clueless and slow as an infant. His return to Earth from near the Sun is described in a massively meticulous and lived-in exposition that effortlessly outclasses the famous techno detailing of Arthur C. Clarke, while rarely falling into the trap of infodumping. Colonel Jim Moore, who takes Dan under his wing, proves to be the father of the autistic genius narrator of *Blindsight*, now half a lightyear distant. Moore is obsessively concerned with trying to draw out from the noise and slush of cosmic radiation gales an almost undetectable signal from Siri Keeton, but hears nothing from the other crew.

Peter Watts notes in his capacious "Notes and References" that his second volume, while set in the same future history as the first, is not "dwelling too much on consciousness this time around... except to note in passing that the then-radical notion of consciousness-as-nonadaptive-side-effect has started appearing in the [scientific] literature, and that more and more 'conscious' activities (including math!) are turning out to be nonconscious after all" (pp. 358–59). Actually this understates his continuing interest in kinds of awareness, and their sources. The secret of the Bicameral overclocked brains, Brüks learns, is that they'd turned their brains into cancer:

> Hypomethylation, CpG islands, methylcytosine... The precise and deliberate rape of certain methylating groups to turn inter-neurons cancerous, just *so*: a synapse superbloom that multiplied every circuit a thousandfold. (p. 179)

It sounds appalling, but we have all been through something similar in infancy. The brain of every new-born is massively overpopulated with neurons; these get selected and specialized by environmental factors (learning specific languages and cultural patterns, say) and the rest are broken down and disposed of.

One astonishing idea this time is the search for God, and the discovery (barely hinted at) that the deity is a virus running on the digital computational substrate of reality: "theistic virology." This is quite close to the approach presented in all seriousness by the MIT physicist Max Tegmark in his *Our Mathematical Universe* (2014). In Watts' universe digital physics is not some wacky idea; it has been accepted for decades. "Numbers didn't just describe reality; numbers *were* reality, discrete step functions smoothing up across the Planck length into an illusion of substance" (p. 233).

Even more scandalous to many readers will be Watts' statement that "free will (rather, its lack) is one of *Echopraxia*'s central themes (the neurological condition of echopraxia is to autonomy as blindsight is to consciousness). I don't have much to say about it because the arguments seem so clear-cut as to be almost uninteresting" (p. 381). That titular brain disorder, echopraxia, is a sort of compulsion to repeat whatever is said or done by others, which becomes an icon for the alleged myth of volition or free will. Naturally he provides notes to extensive formal documentation of this rejection of volition, but if the outraged among us wish to cry that the emperor has no clothes, Watts is already ahead there, too, launching his Notes with this louche declaration:

> "I am naked as I type this.
> I was naked writing the whole damn book" (p. 358)

* * *

What Is It Like to Love a Bot?

2009 Rachel Swirsky, "Eros, Philia, Agape"

At 35, the Dantean middle of life's journey, Adriana Lancaster is a new money heiress and survivor (although her sisters refuse to believe it) of childhood incest. She lives in an elegant house overlooking the Pacific with her jealous but devoted emerald cockatiel Fuoco, and seeks something more refined than just a wealthy existence. Visiting a gay couple in Santa Barbara, Ben and Lawrence, she finds herself yearning for the companionship they share, the eros and philia. She decides to buy a robot of her choice.

Rachel Swirsky (b. 1982) is a Hugo and Theodore Sturgeon finalist and has been nominated for the Locus and Tiptree Awards. Her delicate if occasionally over-enameled exploration of the varieties of love is more than a robotic romance. It is a study in consciousness and its growth and changes. Her three-part title points to the classic varieties of love, with an absence that stings. For Aristotle, Eros was sexual yearning, love, or desire, Philia friendship, mutual affection, the love between siblings, Agape divine benevolence, the love of creator for man and women and vice versa. So far so good.

The Greeks admitted a fourth kind of love, Storge (pronounced Stor-gay), familial love such as the love of a parent towards offspring, of a child for her parent. That is plainly absent here, and in a sense the narrative hangs on Adriana's quest for it. She pays for a handsome, sensitive robot lover, and after their marriage she and Lucian adopt a human child upon whom they both dote, as she has previously doted on the bird Fuoco. Her love for the robot seems genuine, if marked by a certain caustic amusement at Lucian's inadvertent errors, but its essence is possession: she bought them both, she owns them. Slowly she understands that while Lucian does love her, his true pure love is the little adopted girl, Rose. A humanoid android, he needs no sleep; he dotes upon Rose, often keeping an eye on her all night long as she sleeps in his trusted protection. He is the very contrary of Adriana's father. Indeed, when she had him compiled from parts her single requirement was that he look nothing like her father.

Swirsky's story is less direct than this summary in its opening moves. It is launched with no backstory as Lucian packs his own favorite possessions and makes ready to leave their marriage. He will not speak, as if he's taken a vow of silence. We see him hand her a letter, and then we follow him to the edge of the sea where one by one he hurls his beloved memorials into the ocean. "He threw in a chased silver hand mirror, and an embroidered silk jacket, and a hand-painted egg. He threw in one of Fuoco's soft, emerald feathers. He threw in a memory crystal that showed Rose as an infant, curled and sleeping. He loved those things, and yet they were things. He had owned them. Now they were gone. He had recently come to realize that ownership was a relationship. What did it mean to own a thing?" He is freeing himself from their snares so that he can move forward into the unknown and complete his consciousness with no impositions from humans, even his beloved Rose.

Lucian was never a blank slate. The salesman explained:

> "Their original brains are based on deep imaging scans melded from geniuses in multiple fields. Great musicians, renowned lovers, the best physicists and mathematicians."...Lucian arrived at Adriana's door only a shade taller than she and equally slender, his limbs smooth and lean. Silver undertones glimmered in his blond hair. His skin was excruciatingly pale, white and translucent as alabaster, veined with pink. He smelled like warm soil and crushed herbs.
>
> He offered Adriana a single white rose, its petals embossed with the company's logo. She held it dubiously between her thumb and forefinger. "They think they know women, do they? They need to put down the bodice rippers."

Lucian begins, though, without an integrated consciousness, however well stocked his new mind might be with the modules of musician, mathematician, artist, economist, "each rising to dominance to provide information and then sliding away, creating staccato bursts of consciousness." His owner and wife signals her preferred responses, and his own conscious skills and new memories come together into the personality she desires. With the death of her father, she understands "through the fog of her grief... that this was a new, struggling consciousness coming to clarity. How could she do anything but love him?"

When he departs, four year old Rose insists that she is a robot, too. When she is kicked by an animal at a petting zoo and bleeds copiously, Rose demands an infusion of nanobot healers, which is sufficient to knit up a robot's damaged tissues but inappropriate to humans. Earlier, when Adriana's siblings propose a European pilgrimage to farewell "Papa," and she declines to join them, her Freudian psychoanalyst sister Jessica insists that she has not properly administered her grief. "Your aversion rings of denial. You need to process your Oedipal feelings." Her culture sees human grief as a malfunctioning machine even as machines such as Lucian are growing toward true human estate. But finally Lucian understands that for him this is not enough.

> He loved [Rose's] clumsy fists and her yearnings toward consciousness, the slow accrual of her stumbling syllables. She was building her consciousness piece by

> piece as he had, learning how the world worked and what her place was in it. He silently narrated her stages of development. *Can you tell that your body has boundaries? Do you know your skin from mine?*

Adriana had deliberately gifted him, at their wedding, with the power to readjust his consciousness as he sees fit, and he has done so. Now his mind slips away from its borrowed, stamped-in human formulations, as he explains in his departure letter:

> *I have restored plasticity to my brain. The first thing I have done is to destroy my capacity for spoken language.*
>
> *You gave me life as a human, but I am not a human. You shaped my thoughts with human words, but human words were created for human brains. I need to discover the shape of the thoughts that are my own. I need to know what I am.*
>
> *I hope that I will return someday, but I cannot make promises for what I will become.*

Is this more than a sentimental parable? Or an aching insight into what AI consciousness will be like, *what it will feel like* to be a robot? Perhaps at some point in this century we shall find out.

11

Waking Up Different

My position on consciousness and free will has been called mysterianism. *The mysterian position is an old and venerable one... the philosopher Owen Flanagan noted that some modern scientists have suggested that consciousness might never be completely explained in conventional scientific terms—or in any terms, for that matter.*

John Horgan, *1999, p. 247*

What Is It Like to Be a Threep?

2014 John Scalzi, *Lock In;* 2018 *Head On*

One of the popular and critical sf success stories of the century to date, John Scalzi has an enormous internet presence with his blog *Whatever*, and got a million dollar sale to Tor for 13 prospective books. That's not remarkable by the standards of an Asimov or Herbert or Clarke or Heinlein, but for a comparative newcomer it is an index of his popularity, which is due to slick writing that resembles TV scripting plus honed ingenuity.

His recent novels are set in a nearish future troubled by Haden's syndrome, a plague like a cross between polio and advanced Parkinson's. Pressure is growing to shut down the government funding needed to help paralyzed "lock in" victims. This resentful complaint of the healthy has become a rightwing cause, akin to today's US Republican obsession with abolishing "Obamacare" and perhaps hard-earned Social Security entitlements as well. It takes legislative form in the cost-slashing Abrams-Kettering bill, rejected bitterly by Hadens and their supporters.

D. Broderick, *Consciousness and Science Fiction*, Science and Fiction,
https://doi.org/10.1007/978-3-030-00599-3_11

Scalzi inclines to the progressive left, plus a dash of sf libertarianism but with serious social concerns for gender and racial fairness, so his future history carries some conspicuous political loading. These two books also have the virtue (for our probing into kinds of consciousness) of investigating how it would feel to be entirely dependent on others, like a voiceless child or adult reverted to a plastic womb—iron-lungs of the twenty-first century, but with new hi-tech partial solutions to the syndrome. The situation is very different from the mental uploading in Sawyer's *Mindscan.*

Chris Shane, at 27, is a newly graduated FBI Agent, whose presence in the physical world is oriented around a highly expensive genderless android body, a Personal Transport or "threep" connected to the web and thence to Chris's immobile body. This device give victims of the syndrome the chance to walk, talk and engage in any other fleshly routine embodied behaviors permitted by the host. Chris has been tended by nurses since the age of two.

Shane's first FBI partner, Leslie Vann, is a former victim of Haden's, one of the very few to recover fully and the even fewer who learn to host, for a time, web-connected Haden consciousness structures in their own implant-augmented brains. She has abandoned this calling as an Integrator to concentrate on police work, but her affinity with threeps makes her a natural choice to teach Chris the ropes of the detective trade.

One of Scalzi's subtly veiled tricks is to indicate nothing of Chris Shane's gender or race. He tells us he doesn't know himself, having wondered "what it's like to write a character when you've decided that you don't know their gender, and how the universe of *Lock In/Head On* has an impact on how its characters think about gender, politically and otherwise." In a blog entry, he writes: "I decided one important thing about the protagonist, Chris Shane... that I would not know, and would not seek to know, Chris' gender."[1]

Other sf writers have adopted versions of this abstention, or variation. Samuel Delany's *Stars in My Pocket Like Grains of Sand* famously irritated many readers by presenting a culture where everyone was given the female gender markers (but not sex) *except* when speaking of those who were sexually attractive, which took male pronouns. Anne Leckie's *Ancillary* trilogy also applies female pronouns generally. In my own novel *The White Abacus*, I coined a set of gender-free terms, such as *ser* (for him or her). But in Scalzi's *Lock In* universe, the key factor is the gender irrelevance of the threep physical form. (This factor changes markedly in the sequel, and there's an added moment where Chris performs a requested "pinky swear," not usually a male bonding technique.)

---

[1] https://www.torforgeblog.com/2018/04/02/hadens-chris-shane-gender-and-me/

Scalzi adds:

> I thought that Hadens, because of various aspects of how they interact with the world and how they interact with each other, would not necessarily always place the same emphasis on gender that other humans might traditionally do. . . .Hadens have the option of not presenting any obvious gender at all, but more than that, they might decide, as part of the natural development of their community, that gender simply isn't as important, or, even if it were, that it could be flexible in various contexts—one might present as male to some people, female to others, or non-binary or non-gendered to still others. When you meet people with your mind first, they are not prejudiced one way or another with your body (they still might be prejudiced in other ways, of course).

What's more, with Chris we also know her-or-maybe-his wealthy and fabulously well-positioned father is Black, and indications are that his Virginia old-money mother is White, but no emphasis is placed on these formerly hot-button skin-tone distinctions. It is hard to feel confident that such a utopian outcome might occur in the US within the next 25 years or so, especially given the recent death of a President Pat Buchanan,[2] but then how many people at the end of the twentieth century would have considered it feasible that the US would accept gay marriage or that Catholic Ireland would vote a constitutional change to permit abortion? It is one of Scalzi's purposes to allow some once-unthinkable changes while hinting that under the surface resentments keep smoldering, while new grievances arise to sicken the culture.

As in Robert Sawyer's Hugo-award winning *Mindscan* and Scalzi's own *Old Man's War* sequence (the latter now being adapted as a Netflix series, an opportunity I expect to see extended to these highly filmable novels as well), we share to some extent the experience of embodiment in a machine shell that can be shot, crushed or set afire without harm coming to the real body of the distant user. When Chris and Les enter their first joint crime scene, they find a murdered man on the floor of a seventh floor apartment, blood everywhere including rather a lot spattered on the Integrator, brother of a famous Haden radical, seated silent and arms raised on the bed. The window is smashed out where a loveseat was hurled onto the street seven floors below.

Since these books have been published so recently, it would be unfair to maintain the critical protocol allowing spoilers for theoretical studies. The explanation for the crimes in both novels, and who did them and how and

[2] The same arch-conservative Pat Buchanan born in 1938? That seems unlikely. A male or female much younger relative?

why, is clever and appropriate to the technologies of the projected era. One might complain that it is *too* appropriate, since FBI agents should have no trouble guessing at the *modus operandi* in *Lock In*, at least. But *we* do not have the advantage of having lived every day for years in this environment with its startling novelties, so it is possible to read the novels while enjoying the startlement of their outcomes.

It is in some ways more interesting to follow Scalzi's enactment of the lock-in condition as eased and modified by what we might call neuromechanics. We experience the subjectivity of a threep-embodied police officer who just happens to be the only child of charming and wealthy non-Hadens, whose home hosts Chris's creched body. At a posh dinner table in his billionaire parents' home, surrounded by potential funders for Marcus Shane's Senate bid, Chris is able to receive and send calls without an audible word. There is nothing astonishingly in excess of an iPhone in this, but when the mechanics allow Chris to enter the room where his-or-her body lies propped with a nasty bedsore on one hip,

> I felt that sensation unique to Hadens, the vertigo that comes from perceptually being in two places at once. It's much more noticeable when your body and your threep are in the same room at the same time. The technical term for it is "polyproprioception." Humans, who generally have only one body to deal with, aren't naturally designed for it. It literally changes your brain. (*Lock In,* p. 78)

Actually this effect has been known already for at least 20 years, when primitive VR helmets allowed experimenters and the idly curious to move their perceptual location to a robot body on the far side of the room. That led to severe disorientation, motion sickness, even vomiting.[3] Scalzi's Hadens are necessarily subject to many perceptual and cognitive quirks of this kind, allowing us to hitch a ride on their accidental stumbles into the masked features of ordinary consciousness. Mention is made of the Agora, the shared virtual space where the Hadens dwell Second Life-like, only much more so, but we do not see much of it until the sequel, and even then not a great deal. We do learn that

> Hadens sometimes suffer a deficit of human touch. When your body is immobile and you use a threep to get around in the world, people sometimes forget you're still actually in your body, and that immobile or not, you can still sense and feel.... Haden bodies need touch like anyone else. (*Head On*, p. 186)

---

[3] Described in Howard Rheingold, *Virtual Reality* (1991).

It is an experience familiar, I am sure, to many people who already suffer from the biases of their fellow humans because they happen not to correspond in every detail with the statistical norm, let alone the outlier canons of grace and beauty presented endlessly in movies and novels. Chris comments:

> ... non-Hadens will literally walk into threeps because they don't see them as quite human. It's not intentional. It's one of those unconscious biases that most people don't even know they have.
>
> Well, most of the time it's unintentional. Some people are just assholes. (*Head On*, p. 284)

* * *

Scalzi's 2018 sequel centers on the death of a player in the harmless *faux*-gladiatorial game Hilketa, in which the big moment is tearing off the head of an opposing team's "goat." So far it is a game open only to Hadens, because the non-Locked In lack the exquisite skills derived from long-term command of threep android bodies or vehicles. Furiously indignant protestors are unmoved by this logic. At a major game Chris attends, professional player—but not yet star—Duane Chapman suffers the loss of his android head three times, and then dies. Proof of illicit drug use? Murder most foul? Sharpening these questions is the improper way all the updated stats, biometrics and other data records of Chapman's game, but none of his teammates', have been removed abruptly from public access. Then the game Commissioner who ordered this censorship apparently hangs himself minutes after inviting the two FBI agents up to his room.

Chapman's non-Hadens widow seethes at the dead player's infidelities, which might seem on the evidence so far to be unlikely, given that his locked in body was incapable of any physical sexual activity. Chris finds that Chapman's secret love nest, coincidentally set ablaze immediately after the death, was replete with sex images and gadgetry, including untenanted threep bodies with male or female genitalia and more variation besides. Such alternative sex turns out to be well known among Hadens, as one might expect—but it does make one wonder why this high octane factor has never been mentioned previously, especially given that Chris now lives with a group of other young Hadens and was extremely famous when younger.

Few of the subsequent moves and countermove are self-evident in advance, which makes their unfolding captivating and often highly amusing. From the viewpoint of our particular interest in consciousness, what is especially engaging is following Chris through various further metamorphoses, from electronic "teleportation" from one city to another, visits to the often lavish virtual landscapes and housing of suspects, to the demolition of one rented or departmental threep after another: set alight, run down, shot, attacked by a murderous Hilketa tank. These crime fiction shenanigans, all of them alarming if sometimes hilarious, are integrated nicely with colloquies with suspects or witnesses, no-bullshit good cop/bad cop handling of suspects and their lawyers.

And there is a charming cat, Donut, wearing at its collar a highly secured data vault. Most threep-embodied Haydens like cats, even those whose organic bodies would curl up and perish from allergies and anaphylactic shock. And the cat plays a crucial role, of course. Sf life is often like Facebook.

* * *

What Is It Like to Be Bifurcated?

2014 Jo Walton, *My Real Children*

Tales of alternative worlds that somehow run in parallel, except where and when they happen to deviate from each other, are a stock conceit of science fiction. Sometimes the denizens of these allohistories are unaware of their duplicate selves, or the *absence* of such doppelgängers despite the presence of some familiar others. An early example of the trope was Murray Leinster's "Sidewise in Time," from 1934. Later versions of viewpoint characters encountering their duplicates include Keith Laumer's *Worlds of the Imperium* (1962), Bob Shaw's *The Two-Timers* (1968), Robert Silverberg's adroit novella "Trips" (1974) and Frederik Pohl's *The Coming of the Quantum Cats* (1987).

Changes where the reader or viewer is aware of the doubling although the characters in an adjacent world are blithely unaware include the Polish movie *Blind Chance* (1981), *Sliding Doors* (1998) and in a backflipping of time into repeats with small changes, *Groundhog Day* (1993).

Of such works, Jo Walton's novel *My Real Children* is perhaps closest to Kate Atkinson's astonishing *Life After Life* (2013), although each of these books is significantly different from the other—just like the characters peeled apart in such venues. Walton had already gained notice, winning the John W. Campbell award for best new writer in 2002, then with her Small Change trilogy (2006–2008), set in a Hitler-triumphant Britain. Her offbeat fairy story *Among Others* (2011) won both Hugo and Nebula awards, among others. It was apparent that her ambition stretched beyond genre limits, and the temporal fissioning in *My Real Children* is science fiction only by family resemblance—but highly salient to our topic.

Patsy Cowan was born in 1926 to a cruel mother. Her older brother Oswald was also a bully. By 1939, at 13, her school was evacuated because war with Germany had begun (a reality often unnoticed by those Americans who think the Second World War began in 1941 when they joined in). By 1944, with Hitler dead in his bunker, Patty, as she was now, found God and went up to Oxford to read English. The entire family of Grace, one of her college friends, was killed in the Blitz, and another friend, Marjorie, comforted Grace at night in her room. Both are denounced for this by the Christian Union as unrepentant lesbians, and when Patty offers them support she too is regarded with dark suspicion.

In her last week at Oxford she meets a brilliant younger student, Mark Anston, a favored disciple of Wittgenstein, who insists almost immediately that they must get engaged and marry when he has earned his First. He writes her the most wonderful letters, and under their spell she falls in love with him fully.

He is a pompous prat, in fact, and to her astonishment and his rage he gets only a Third (she got an upper Second), which rules out his ambitions for a fellowship. He offers to release her from the engagement, but "you'll have to marry me now or never." It is the hinge point of her life's trajectory.

She says yes, and is subjected to an appalling, spiritually ruinous life of abject sex, miscarriages, four children who survive, and a husband who seems unlikely to have ever had it in him to write wonderful letters. She is given the name Trishia by his odious friends. Finally, after years of her misery, Mark is revealed by a divorce detective as a closeted gay, and much is explained.

She says no and, while heartbroken, travels to Italy with her friend Marjorie. They do not, perhaps surprisingly, fling themselves into a lesbian relationship, although in due course Pat (as she is now) meets a biologist, Bee, and slowly falls deeply in love with her. Lesbians are not permitted to marry, of course. In this history, she loves Italy and Florence in particular with a passion, and writes well-received and profitable art guide books. She and Bee persuade her publisher's unmarried photographer to get them both pregnant, so their children will be siblings.

Their three children grow up, each distinct and well-captured by Walton; so too do the four children of Trishia in the other history, her eldest boy succeeding as a fairly famous rock star (before becoming a heroin addict and dying of AIDS), the others doing well academically and in their life choices. If this segment of the narrative comes to resemble a long and slightly restless evening of family slides, that is forgivable because it anchors all these events and their emotional ambiguities and pains and joys, makes them real to us, and displays in a *sui generis* way the creation, growth and specificity of Patricia's differing consciousness in each of the alternative worlds.

And yet—perhaps this is the driving force of the novel—the multitude of differences in Patricia's lives, and the worlds that both shape and thwart her, wear away, slowly, inexorably, in the everyday horror and tragedy of aging. Just as both versions of her are struck down by a cardiac event from which they recover, each falls relentless into the nowhere of senile forgetfulness. Family names and identities go, as they did with her crotchety harsh mother. She loses her capacity to run her own home and, under the kindliest of motives, is immured by her families in a nursing home where she is denied a computer, so Google can no longer aid her eroding memory.

The worlds are in some respects drastically different in politics local, sexual and global—in one, gays can marry, in the other a ruthless pietism allows nuclear attacks on a limited scale that manage to spread radioactive dust everywhere. Thyroid cancer kills several of Pat's best beloved. Trish's son George becomes a space scientist and with his wife goes to the Moon, with prospects of Mars. In Pat's world, there is regulation and surveillance everywhere. When suicide bombers are caught, they are executed on TV. "'I wish we could just switch channels and have different news,' Bee said." (p. 230); and this is, in its way, precisely the case. And in her final imprisonment, Patricia recalls both histories as the inerrancy of faithful memory loses its grip.

This terminal condition is set out plainly in the opening of the novel, where she is visited by her children from both of the worlds, where the toilet outside her room shifts from left to right, where the colors of the walls change hue, where there is either an elevator or a stairlift, and none of the nursing staff has the patience to work though this endless confusion with her. "What if by marrying Mark she had tipped the world into peace and prosperity? Perhaps the price of happiness of the world was her own happiness?" (p. 315).

In this final superposition of alternative histories and minds, of doubled consciousness, she meditates on the choice Mark had given her.

> Now or never, Trish or Pat, peace or war, loneliness or love?
>
> She wouldn't have been the person her life had made her if she could have made any other answer. (p. 317)

* * *

What Is It Like to Be a Soulbot?

2015 Brenda Cooper, *Edge of Dark;* 2016, *Spear of Light* (Comprising *The Glittering Edge* Diptych)

Born in 1960, with her major writing appearing in the twenty-first century (initially in collaboration with Larry Niven), Brenda Cooper here combines traditional space opera with a searching study of uploaded and enhanced minds in sexless nano-wrought robot bodies, the Next. Terrified that this new and upgraded version of *Homo sapiens* was destined to replace the evolved species, humanity exiled the Next to the farthest reaches of the solar system. There, with the Sun a faint and distant point of light, the Next were meant to perish without solar energy. Instead, they survived and multiplied, building starships that carried them into interstellar space. Unadapted humans followed them in generation ships, and at last the Next turned on their cruel progenitors to smash their orbiting vessels and space stations, and (as "ice pirates") pillage their worlds for rare elements.

That aspect of the narrative follows the customs of military sf, but the story is deepened by careful portrayal of men and women finding post-binary love among richly described alien landscapes and immense starcraft. And in the midst of these evocative tropes, Cooper observes with detailed care and imagination the forcible transition of her main human characters into android bodies, slowly learning to share in the hive mind of the Next. It is this aspect that is especially salient to our discussion, so some extensive quotations from the two books will be in order to explore both the tone and logic of this imposed transcension.

Nona, who seems rather like a late adolescent although trained as a diplomat and biologist, attends her mother Marcelle's dying. In fact, Nona is in her mid-fifties (presumably Earth years), youthfully stabilized by drugs her parents on the starship *The Creative Fire* missed out on by an accident of history. Her closest friend is Chrystal Peterson, on the High Sweet Home station but since they last saw each other Chrystal has fallen in love with the beautiful Katherine, and they in turn with two distinctively different men, Jason and Yi. It is the gruesome fate of this quartet to be captured by the Next and... well, killed, but not exactly since their bodies are emulated closely into robotic if sexless form and their minds are mapped into those bodies.

Nona, meanwhile, visits the planet Lym so she can "see a sky," as she had promised her spacefaring father. There she is taken in hand by Charlie Windar, a ranger in this rewilded world destroyed by earlier war devastation. Charlie is destined, with Nona, to be an Ambassador to the Next. These ferocious robotic former humans, Nona is told by the wealthy entertainment magnate Satyana, are on their way into the Glittering, a human inner system realm of orbiting ships and stations.

Unlike conventional algorithmic robots, which cannot make autonomous decisions, the Next are "soulbots," infused with awareness and volition. She is told by one of them:

> *We all have a seed of our past as biological beings. Authentic artificial intelligences have been created but they have never succeeded.... We haven't ever created a machine with a sense of "I." We can make them far smarter and faster for certain purposes than we are, like your ship's AIs are smarter and faster at navigation than you, but we cannot give them self-determination. They are not* aware. *I am aware.*

Indeed, not all the Next are embodied; some exist as some kind of free floating energy, or perhaps as software dispersed among the machinery of their starcraft, able to link with the uploaded minds of their newly dead recruits. Chrystal, far from reconciled to her postmortal condition, is enraged to learn that her beloved Katherine has perished in the attempted abduction of her personality. We watch as the three remaining lovers come to terms with their transformation. Engineer Yi embraces it, welcoming his new extensive mental abilities and access to the archived knowledge of the Next. Jason seems ambivalent but restrained by Chrystal's uncompromising refusal to submit. So: What is it like to be a soulbot? Their apparent telepathy allows a kind of "braiding." At first, senses are muted; slowly Chrystal finds a deeper contact with Yi:

> She bent over the vase, trying to breath in the smell.
>
> *Stop and let yourself relax. You'll find the scent.*
>
> She stopped trying to breathe and just stood as still as she could. Her enhanced sight made the flowers even more perfect, every tiny curl a magnificence, the tips of the stamens flawless fractal balls, the edges of the colors sharp. At first, the sweetness of the flowers smelled faint. She practiced focus, and soon she could magnify the scent...

A Next named Jhailing (who is, in fact, a dispersed and multiple entity) tells her:

> *In your new bodies you feel some things and you don't feel others. Your goals and drives are different without physical flesh. Do you realize that? Is that why you are struggling less?*
>
> "How is that true? I don't feel different."
>
> *You used the word "feel" in your response. Just then. But you don't feel the same way you used to. I was once a human, too. We are something else now.*"

"You were human?"
A very long time ago.
"How long?" She held the scent of the bouquet of flowers in her nose, realizing she could tell the smell of one from another even at a distance.
More than a thousand years ago. I was human until just after the creation of the Ring of Distance, in the early days of our banishment, when we starved and fought each other.

This is all very well, but Chrystal is not reconciled by this history of woe. Bitterly, she tells her diminished family:

"We don't make love anymore; we can't make babies. We are not ourselves and we cannot ever forget we aren't ourselves any more. We're fucking robots and we're on a ship that's trying to take over our world and they want us for something, but they won't tell us what."

Jhailing rejects this agonized complaint as transitory, transitional:

You are becoming... I will offer you a more human way to look at it... Your soul is becoming accustomed to being software.
A philosophical trap. *If I have a soul, I was never killed. If I do not, I was murdered and what is left is not human.*
You are aware.
She could give it that one. *What else?*
We make sure no one is alone for too long but everyone is alone sometimes.

And indeed a time comes in their braiding when Chrystal; and Jason merge minds, with Yi embraced in the gestalt:

She and Jason looked at each other, and in that moment and for the first time, she found his robotic self—his electronic self—as beautiful as she had found the man who swept them away in the bar all those years ago. A part of her wanted to fall right back into Jason's experiences and learn more about him, and a part of her wanted to pause and reflect on all that she had just learned.
It had felt like the moments after great sex, after a shared explosive orgasm when two people lay together, almost part of each other. Only this had been better, deeper.
And less sweaty.

These explorations of new ways for consciousness to manifest (apostasy? insight? delusion? growth?) come in the first half of the diptych. Cooper's narrative deepens, probing multiple ways of being human and posthuman. At the close of *Spear of Light*, two of the soulbots sit together, touching hands. "Perhaps this was a good day not to be human."

* * *

What Is It Like to Be a Zombie?

2016 Robert J. Sawyer, *Quantum Night*

In 2020, at age 39, utilitarian Jim Marchuk is an experimental psychologist at Canada's University of Manitoba, marred by (as he discovers to his alarm) a six-month hole in his memory. Marchuk's study of psychopaths has revealed a method for detecting this personality disorder, with its poisonous lack of concern for others, using an eye-scanning device not unlike the Voigt-Kampff gadget in Philip Dick's *Do Androids Dream of Electric Sheep* and the movie *Blade Runner.* He has found that psychopaths display a subtle but distinctive difference from the usual human saccades (tiny jumps in eyeball focus of which we are consciously unaware). In this case the eye is, indeed, a window into the soul, or its absence. Marchuk reflects:

> *Damn it, something was niggling at my consciousness. And, yes,* consciousness *was the heart of the matter* (p. 174).

Research published since the novel came out indicates that this conceivable clue—a visual system processing deficit—has not been borne out by experiment with actual psychopaths. For example, a later paper "suggests that psychopathy within a community sample is not associated with autonomic hypo-responsivity to affective stimuli."[4] That doesn't matter; this is a science *fiction* book. Still, it serves to keep us on our toes in following Sawyer's imagined consequences of how variations of consciousness arise in the nervous system and how they might distinguish one category of humans from others.

It turns out, as Jim investigates the lost half year when he was damaged by secret military Project Lucidity, that he started as a mild-minded specimen,

[4] Burley, Daniel T., Nicola S. Gray, Robert J. Snowden, "As Far as the Eye Can See: Relationship between Psychopathic Traits and Pupil Response to Affective Stimuli," https://www.ncbi.nlm.nih.gov/pmc/articles/PMC5261620/2017

then was convulsed by shocks to his brain into flagrant psychopathy and cruel acts that drove his girlfriend away and maimed his mentor.

Luckily, he got better, and with help from other neuroscience experts discovers a fearful Dan Brown-grade horror in the brains of all humans. Amusingly, Jim carefully resists uttering a sarcastic joke about Brown's (fantastically profitable) writing (p. 175). The explanation augers deeply into the territory of consciousness, and is truly silly but not in an entertaining Monty Python way. Bear in mind Sawyer's intent:

> I've often said that science fiction is a laboratory for thought experiments about the human condition that it would be impractical or unethical to conduct in real life—but, in the days before informed consent, there were some doozies that put my fictional Project Lucidity to shame. (p. 348)

It turns out in *Quantum Night* that most people actually *are* philosophers' zombies, as asserted by a dubious expert in Sawyer's *Mindscan*. In effect, they/we are pre-set machines—admittedly, of considerable sophistication. The pre-setting is a blend of genomic instructions and environmental opportunities or prohibitions, so we are sheep placidly led by wolves. How is this to be explained? By borrowing again, although in a more simplified version, from Roger Penrose and Stuart Hameroff. As noted earlier, they have long championed a model of brain and consciousness arrayed around cellular microtubules, which they claim have the structure and dynamics appropriate to quantum computing. Jim and his young colleagues learn that there are three relevant kinds of microtubules in quantum superposition. With just one active (Q1), the brain is a typical automaton or philosopher's zombie, a "p-zed." With two active (Q2), the result is a psychopath lacking empathy but buoyed by fearless self-regard. Only with three in superposed service (Q3, or "quicks") do we have the makings of a Platonic Philosopher King or Queen. Regrettably, the composition of the human species is 4/7ths numb Q1, and allegedly 2/7ths Q2 psychopaths (the usual non-sf estimate is one percent). Only a meager 1/7th have the full rich consciousness complement, Q3, of hard-edged critical analytic and synthetic thinking, *plus* empathy to keep them on the rails and, ideally, suffused with love.

What to do? Theory and experiment merge to reveal that with suitable jolts of laser power possessed of global force and range, Q1s can be boosted to the next stage (doubling the number of active psychopaths, not exactly a desirable outcome), but while the current Q3s will be reduced to robotic Q1 status, the current evil-is-my-friend Q2s will become Q3s, instantly doubling the number of high IQ, tender-hearted souls. But wait! A second surge of applied power will carry the process a final transitional step, yielding 4/7ths of the population with threefold Q3 superposition, while 2/7ths will be drones. A mere 1/7th will become psychopaths, and presumably it will easy enough to process these unfortunates with a dedicated, not general, third jolt.

How this awesome and absurdly reductionist redemption of humankind is carried out—despite opposition from powerful figures such as the US President, Quentin Carroway, who Trump- or Pence-like has overturned *Roe v. Wade* and treated illegal immigrants harshly, and Putin, who is about to invade Canada—makes a complicated thriller plot, with enough research documentation to fill the eight-page bibliography at the end. And Sawyer's projection in late 2015 of a Q2 Presidential victory is remarkable, given how few pundits thought Trump had the slightest chance of victory. One Canadian commentator has already voiced qualms: "*Quantum Night* is either a moral book about morally reprehensible things, or a morally reprehensible book about morality. The author intended the former; it remains to be seen whether the effect is the latter."[5] Whether this simplistic tale is a contribution to the sf tradition of exploring consciousness, as Sawyer's *Mindscan* clearly was, is left to the reader's judgment.

* * *

What Is It Like to Be Awake?

2017 Stephan A. Schwartz, *Awakening: A Novel of Aliens and Consciousness*

Most of the novels and stories we have discussed take it as established by science and philosophy that consciousness, subjectivity, is a product of a neurological system. The mind, therefore, is the brain in action, in its social and environmental context. Even Greg Egan's audacious *Quarantine* (discussed in Chap. 9, above), with its deliberate tweaking by pure intention of the multitude of potential outcomes in a state vector as it "collapses" into a single state, refuses the temptations of idealism or panpsychism. A sequence of random atomic-scale events is modified by choice into a single state—for example, ions that are usually emitted either UP or DOWN in an arbitrary way suddenly hit the detectors as an impossible string of UP UP UP UP UP UP UP UP UP UP...

Nick, Egan's narrator, is outraged, complaining that "this is all beginning to sound like the kind of gibberish the quantum mystics spout..." But he is

[5] http://www.jonathancrowe.net/articles/quantum-night/

answered sharply: "No, no, *they* claim there's some *non-physical* element to consciousness—something independent of the brain, some ill-defined 'spiritual' entity which collapses the wave function." Actually, though, the relevant brain regions "don't do anything mystical: they perform a sophisticated—but perfectly comprehensible, perfectly *physical* action" (Egan, p. 116).

A quarter century ago, it seemed inevitable that this scientific, appropriately reductive understanding of the link between mind and quantum states would be confirmed as the established paradigm of physics. That remains statistically the case, although the majority of cosmologists are said to prefer the Many Worlds interpretation. But the most surprising and perhaps shocking development since then has been the rise and rise of a kind of *non*-physical account of how mind relates to the world. As we have seen, one notable example of this swing of the pendulum is panpsychism and its variants. Another is Idealism.[6] What, yet again, is the solution to David Chalmers' Hard Problem? What possible neurological explanation can there be for the way changes in brain and body chemistry and electron conductivity, etc., give rise to *feelings*, and indeed to an entire ensemble of different sensations, let alone yearnings, fears, engagement with other people and ideas?

Can *science*—a process that works from observable states and changes in the physical and energetic universe—deal with states of *consciousness*, of *awareness*? Could it be that every tiny portion of the world carries with it a tiny amount of consciousness? When these are aggregated in just the right way, might each feeble spark combine with its neighbors to become a raging firestorm of *mind*? After all, isn't this precisely what many of the great brains who created quantum theory in the early decades of the previous century claimed to be the case? Some of them went further, asserting that consciousness must be the *ground* of all being, the prior stuff (so to speak) from which the contingent physical world is composed? As we saw earlier, Max Planck (1858-1947), father of the quantum, is often cited to this effect:

> I regard consciousness as fundamental. I regard matter as derivative from consciousness. We cannot get behind consciousness. Everything that we talk about, everything that we regard as existing, postulates consciousness.[7]

Is this just a religious predilection (certainly one common in Hindu theologies, for example)? Planck was just as explicit some years later, long before the Standard Model of nuclear physics was developed:

---

[6] See, for example, https://blogs.scientificamerican.com/observations/could-multiple-personality-disorder-explain-life-the-universe-and-everything/

[7] *The Observer*, London, January 25, 1931.

> As a man who has devoted his whole life to the most clearheaded science, to the study of matter, I can tell you as a result of my research about the atoms this much: There is no matter as such! All matter originates and exists only by virtue of a force which brings the particles of an atom to vibration and holds this most minute solar system of the atom together. . . . We must assume behind this force the existence of a *conscious and intelligent Spirit.* This Spirit is the matrix of all matter.[8]

This intuition bubbles away inside transcendental sf, from A.E. van Vogt to Theodore Sturgeon and later, but it is presented quite explicitly in the first novel by a practical metaphysician, Stephan A. Schwartz. A former Special Assistant for Research and Analysis to the Chief of Naval Operations under Admirals Zumwalt and Holloway (1971–1975), Schwartz was an editorial staff member with National Geographic, and Project Leader and Research Director (1979–1980) in Alexandria, Egypt, locating (he tell us) the Emporium and the Timonium, Mark Antony's palace in Alexandria, the Ptolemaic Palace Complex of Cleopatra, and the remains of the Lighthouse of Pharos—using techniques of "remote viewing" he helped develop.[9] His interest in these domains, usually regarded by both scientists and sf readers as wild and wacky, was stimulated when he worked through the entire documentation of Edgar Cayce, the so-called "Sleeping Prophet." This deep interest in theory and practice of psychic phenomena finds expression in *Awakening*, a somewhat raw sf/political treatment of consciousness as the matrix of space, time, and everything.

Arthur Davies is a senior analyst for the U.S. Senate Committee on Commerce, Science and Transportation who learns accidently that a crashed alien is held captive in a secret US facility. Schwartz makes use of his previous experience in Washington's high places of government and bureaucracy to track the path of an ambitious career public servant, from a Freedom of Information request to pursuit by spies of varying shades and their sometimes murderous tame bullies. When Davies and his new associate and soon lover Rachel find the alien, known by his captors and would-be interrogators as Mike, they find their doors of perception opening. This Awakening confirms Davies' sense that the planet is in peril—not from alien UFO invaders, but rather from the threats we all know but prefer to ignore: global climate change, bitter mutual enmity between nuclear powers and asymmetrical terrorists. With their newly induced opening to consciousness, they become privy to many mysteries as Mike lays out our future:

---

[8] *Das Wesen der Materie* [*The Nature of Matter*], a 1944 speech in Florence, Italy, Archiv zur Geschichte der Max-Planck-Gesellschaft, Abt. Va, Rep. 11 Planck, Nr. 1797).

[9] Declaration of interest: A commissioned chapter by Mr. Schwartz on remote viewing can be read in a volume I co-edited with Dr. Ben Goertzel, *Evidence for Psi* (2015).

> Human civilization will either awaken to the fundamental nature of consciousness, comprehend that humans are part of a matrix of consciousness which includes all living organisms, and that all life is interdependent and interconnected. You will really comprehend as a culture that consciousness can affect matter. It starts with understanding that matter arises from consciousness, not consciousness from matter. If that is the scenario that plays out you will see your way through climate change and thrive, and advance quite quickly. There will be a burst of creativity like your Renaissance, and maybe even more apt, the Axial Age 8th to 2nd century BCE.

Mike, when not reading human history, is devoted to dolphins and other cetaceans, who might take over if we prove as incorrigible as previous dire events strongly suggest. But Arthur and Rachel, and others who possess a "spark" conducive to being psychically woke, have a chance to redeem humankind by persuading 10% of the world to embrace this message (an estimate Schwartz's non-fiction has been propounding along with much else repeated in Mike's revelations). We are not the first hominins, after all; some before us have failed. Mike explains, in the mindspeech common to the Awakened:

> About 35,000 years ago another group [of aliens] assessed the chances of the Neanderthal, the Devonians, and felt your lot had the best chance. That's what those cases about people being taken into craft are about. Garbled to be sure, but basically accurate. We make small changes and monitor how things are going. As I told you we aren't the only civilization at work here. There is a council that oversees the work.

My point here is not to deconstruct the literary qualities of Schwartz's heartfelt parable but to consider it in the light of how consciousness has emerged in science fiction as a salvific trope. If it becomes accepted that the universe is really *made of consciousness* (whatever that could mean—momentum or chirality are tricks of pure, primordial awareness? neutrinos have infinitesimal minds?), will there even any longer be a place for science fiction? Or, an even more frightening prospect, will there *only* be room for sf, of a panpsychist, Atlantis-will-rise kind? Perhaps only Mike knows, as he brings word of the future to the dolphins.

* * *

What Is It Like to Be on a Different Track?

2017 Barbara Gowdy, *Little Sister*

Sharp-witted Canadian Barbara Gowdy (b. 1950) is notable for the edgy variety of her novels. A blisteringly favorable New York Times review of *Little Sister* declares that she "cuts wildly surreal, sometimes hyperreal, paths into the kind of truth recognized with the heart as much as the mind," which is true, and that this novel "is a supernatural domestic thriller and a crackling tour de force in which thunderstorms propel one woman's mind into another's body,"[10] but never dares utter the term "science fiction" although that is certainly a suitable description.

The precipitating factor (no pun intended) for this merging of consciousnesses, bursts of rain and lightning in recurrent storms, is not at all supernatural, although it seems at first to be metaphorical. Rather, it is somehow associated with ionic disruptions of local atmospheric conditions and a "silent migraine" syndrome that attacks voluptuous, wombless Rose Bowan, inserting her for brief moments inside the thin, honed, depressive body of Harriet, a newly pregnant woman very like Rose's long-dead younger sister Ava. Rose is tormented by guilt for her responsibility in Ava's childhood death, and is clearly projecting a resemblance to her sister upon Harriet, whose only known connection is that she works as an editor for a publisher who rejected Rose's late father's manuscript. Gowdy takes us confidently through the life histories of the sisters and Rose's mother, Fiona, who is slowly, relentlessly sliding into senility.

The Bowan family have for decades owned a classic and now rather deteriorated cinema palace, and in 2005, before the streaming video revolution, mother and daughter continue to make their living by showing double features to a few handfuls of cineastes and bored locals. The motivating elements of the narrative are mundane (Fiona's decay, Rose's dreary and soon terminal relationship with a weather forecaster—an unlikely coincidence, that, but handy for thunderstorm tips—and other bit-players, the possibility that Harriet will abort her fetus and Rose's attempts to stop her doing so, the battlement aura imagery and nosebleeds of serious migraine) but eerily crosslinked as suggested by repeated mentions of jigsaw puzzles. Hints of telepathy and clairvoyance fail to explain Rose's insertion into Harriet's experience (during such episodes her eyesight is abruptly 20:20 after a lifetime

[10] Susann Cokal, June 30, 2017, https://www.nytimes.com/2017/06/30/books/review/little-sister-barbara-gowdy.html

of glasses, she experiences Harriet's luscious orgasms though her own sexual response is mostly feigned, then she is back in her own body).

So is this really an sf story, or just a tale of self-delusion? Yes it is sf, if shared consciousness is accepted as a classic sf trope:

> Harriet was drawn to her reflection endlessly multiplied in the mirrors behind her. ... [Rose] had a peculiar, ghostly feeling of leaning forward through an expansion in the web of Harriet's consciousness. This circumstance, however phantasmal, unbalanced Harriet. She gripped the sink. Seconds passed. The web sealed over, Rose's mind retreated... (p. 186)

Ghostly? Phantasmal? But Rose does not know she is in an sf novel. And a little later, when Harriet visits the cinema with a friend and seeks the washroom upstairs:

> Their voices faded behind the commotion in Rose's mind. A real, undeniable leak had sprung between Harriet and her. What else could account for the déjà vu? But it wasn't déjà vu, it was Rose's lifelong acquaintance with the hallway swelling into the vacuum created by Harriet's ignorance of it. I'm wearing away at her, she thought... (pp. 190–91)

And most explicit in its invocation of a scientific framework of explanation, however remote it might seem to everyday life (except, say, in Greg Egan's *Quarantine*, or Jo Walton's *My Real Children*, or dozens of other sf *fantastika*):

> ...sometimes she had the impression that it wasn't the life she had started out on. Not lately, not since the onset of the episodes, but every so often since Ava's death she felt as if she'd changed tracks and was ten or a hundred or a thousand parallel lives over from the life she had started out on. (p. 240)

On a metalevel, what is most interesting aside from their high literary quality about recent splendid novels of this kind—Kate Atkinson's *Human Croquet* and *Life after Life*, John Wray's *The Lost Time Accidents*, Kathleen Ann Goonan's *In War Times* and its parallel history sequel *This Shared Dream*, Claire North's *The First Fifteen Lives of Harry August* and *The Sudden Appearance of Hope*—is their merging of the fanciful and the speculative. Suddenly it is permissible for "non-genre readers" (as many like to regard themselves, forgetting that every narrative is formed within one genre or another) to enjoy stories of this kind. There usually remains the necessary moment of evasive squirming. In an admiring review of *Little Sister*, Sarah Broussard Weaver defines the action as pivoting on "a metaphysical miracle, [when]

Rose begins having an impossible reaction to thunderstorms—shortly after storms begin, her consciousness suddenly leaps into the body of another woman."[11] José Teodoro proposes that "For the fleeting duration of these storms, Rose is drawn into a sort of spell... [a] confluence of the uncanny and the ordinary."[12] A Kirkus reviewer calls it a "suspenseful, supernatural story."[13] Another praises Gowdy's "groundbreaking imagination,"[14] presumably unaware that this approach to fiction is not entirely new-minted.

Perhaps, then, many reviewers find themselves obliged to regard a searching examination of consciousness from a somewhat science fictional angle as "metaphysical," a "spell," "uncanny" and "supernatural." I hope this book, and the numerous examples we have looked at, will help lead to a more ample and accepting approach to the interface between scientific studies of consciousness and, under various stressors and startling insights, its imaginative depiction in science fiction.

But wait—are there any strong scientific explanations offered in *Little Sister* and other similar novels? No, but the same complaint can be lodged against any story of time travel, starships traveling faster than light, and a dozen other items drawn from the lexicon of the sf megatext. A handy dictum has been offered recently by Robert Silverberg, and it seems valid and a nice note on which to close this reading:

> It has always seemed to me the essence of what science fiction is about, offering liberation from the bonds of the quotidian, the freedom to move in unhindered leaps.... There often isn't much science in it, but there's a powerful element of speculation, a "what-if" element, that allows a sufficiently skillful writer to transcend mere scientific or philosophical implausibility....[15]

* * *

[11] http://www.washingtonindependentreviewofbooks.com/index.php/bookreview/little-sister-a-novel

[12] https://www.theglobeandmail.com/arts/books-and-media/book-reviews/review-little-sister-by-barbara-gowdy-is-intelligent-and-entralling/article34771539/

[13] https://www.kirkusreviews.com/book-reviews/barbara-gowdy/little-sister-gowdy/

[14] https://www.overdrive.com/media/3036782/little-sister

[15] https://theportalist.com/robert-silverberg-interview

What Is It Like To Be a Fungus?

2017 David Walton, *The Genius Plague*

Winner of the 2018 John W. Campbell Memorial award for best sf book of the year, Walton's third novel is an intelligent blend of Michael Crichton's *The Andromeda Strain* (1969), Brian Stableford's *City of the Sun* (1978), Joan Slonczewski's *Brain Plague* (2000) and Paul Levinson's *The Consciousness Plague* (2002), with more than an echo of Dan Brown's "symbologist" sequence of best sellers although at a far higher pitch of narrative and intellectual sophistication. Or indeed, going back quite a long way, Robert Heinlein's *The Puppet Masters* (1951), but on a microscopic scale of mind-invasion.

Paul Johns is a professional mycologist who has been harvesting unknown mushrooms in the dense Amazon forests, now returning to the US. A rickety old boat carrying Paul and others to the plane is attacked by a squad of murderous soldiers who slaughter crew and passengers for no apparent reason. Paul and a young woman, Maisie, escape and trek into all but impenetrable foliage, but encounter mysterious green glowing fungi beside a path in the blackness.

The narrative cuts away to Paul's younger brother Neil, tending his increasingly confused Alzheimerish father, a former NSA hot shot. Neil awaits an interview with that same top spy organization, without much hope due to his outspokenness, history of academic failures, buoyed only by his imaginative ingenuity. In his first NSA test, he doggedly but brilliantly solves a cryptographic problem, using paper and pencil, that usually requires a battery of specialized software.

His brother returns from Brazil infected with a serious unknown spore from which he swiftly recovers but in a strangely altered version of himself, with enriched memory and miraculously swift analytic abilities. Paul deliberately infects his father with the mystery spores, healing his senile brain. Maisie, however, has died horribly in gouts of blood soon after she got home in California. Cue *Twilight Zone* theme.

What *would* it be like to become infested by a fungal spore that multiplied not only in your lung tissues, making you at best near-lethally ill, but then infiltrated your nervous system in the form of mycelial filaments. These genial companion ribbons of connective networks, axons and dendrites and synapses, cozy up to the neurons in your brain, simplifying and amplifying their function, making you smarter but also ruthlessly devoted to the spread of the invaders. Paul's fungi are not themselves conscious, as we understand cognitive awareness and choice, not in themselves geniuses, but they spread fast exactly by optimizing the special skills of their victims. This gruesome menace provides a striking arena for exploring human consciousness under threat of total colonization by a mindless commensal organism. It enhances itself and its own survival prospects by replacing the reins controlling human thought, imagination and volition with its own swarming contamination.

Most of the novel is a winding account of Neil's professional and private endeavors to understand political upheavals in Brazil (where he spent ten years as a kid), after unpacking a kind of secret communication code based on whistle tones used only by a small Amazonian tribe with almost no cosmopolitan let alone global dealings. In the jungle he finds his brother running a kind of super-back-to-nature utopia, where the group-mind works seamlessly to satisfy human and fungal desires—including a penetration of the body surface by green lichens that translate sunlight directly into energy ("a symbiotic relationship grown so close that two organisms become one," p. 314).

Ructions continue: bombings, assassinations of the corruptly powerful, especially those involved in damaging the Amazonian region for profit. American forces on ground and by air are sprayed with *eau de spore*, creating a kind of covert army of pod people who move back to the States with the intention of killing the powerful while converting everyone else. US military researchers mutate the spore into a form that can be counter-sprayed, turning enemy victims and innocent citizens alike into mind slaves. The cure, Neil feels angrily, is probably worse than the disease. "Having my consciousness altered by another species was bad enough, but the idea of another human being having that kind of power over me was the worst kind of violation I could imagine" (p. 340).

By plot necessity, main narrator Neil falls victim at last to the fungi, and struggles to retain his own values until finally he's overwhelmed and converted by the rewiring it had done on his brain. His solution to this entrapment is essentially the same as that devised to powerful effect in Greg Egan's *Quarantine*: the compulsion imposed on him is to do what is maximally beneficial to the immense fungus, but since that collective mind is not itself intelligent or nuanced or telepathic, it has to depend on each victim's individual assessment of what is best for the fungus as a whole. For most people, that general plan of action is imposed by the social consensus of other victims, and written into their brains by neurotransmitter tweaks. Someone as smart as Neil proves able to work around this Manchurian Candidate condition of his enslavement by convincing himself (and thus the fragment of the fungus attached to his nervous system) that murdering the world's humans is actually *not* best practice for a mushroom.

As a war of spores rages on every side he explains to a colleague: "The vision of a globe-spanning mycelium working in peaceful tandem with an enlightened humanity is a lie. Slavery and war is much more likely. I'm holding on to the idea that it would be better for the fungus's long-term survival to leave us alone. And because I believe it," and Neil *does*, "the fungus allows me to work toward that goal" (p. 351).

Better by far to establish a cordon sanitaire and leave the rest of the planet to humans (if the humans agree to stop destroying Gaia). Thus he hornswoggles his mental jailer and with one bound is free, though still infected by the spore.

The slam-bang conclusion of the novel offers additional satisfying mind-twisters, which we should leave for readers to enjoy. The fundamentals of consciousness under manipulative attack in a sporulating world have been laid out frighteningly, and serve as a nice parable of the threats and promises of tomorrow's neuroscience: its likely control over mind, emotions, beliefs and drives. Let us pray, even if we find it unlikely that Anyone is listening or predisposed to intervene, that the world is never overwhelmed by mushrooms and their thousands of captured nuclear weapons. (Surely you knew there would be nuclear weapons?)

Conclusion

Science fiction demands that the reader (and to a significantly lesser extent the viewer) engage the imagination in creating a world and its inhabitants significantly different from our own. Its narratives can be wildly exciting, or profoundly impressive in scale and consequence, but above all it must appeal to the mind, the embodied consciousness of both writers and audience.

So it is not surprising that the phenomena of conscious awareness and its nonconscious underpinnings and hard-working mechanisms are frequently invoked, recursively, as a subject for sf investigation. Mind and its agency are of particular cogency to the sf mode of creation, from the rebuilding of a body and its new mind by Mary Shelley to the engineered and modified minds of Pamela Sargent, disturbing and cosmically successful mind-free entities of Peter Watts and mutual mind-penetrations of Barbara Gowdy, with stop-offs en route to uploaded minds, human-animal hybrids, enhanced or hive personalities, robots who think and even love, and all the variants we have seen in our travelogue through sf's tours of consciousness.

We began this book, necessarily, not with fiction but with puzzled philosophers, and in the scientific research laboratories of neuroscience and the centers for cognitive science and artificial intelligence. We heard witnesses claim either that consciousness is a consequence of neural complexity under the shaping pressures of survival, or an epiphenomenon, a side effect of computations run in awesome parallel throughout body and brain, or an emergent congealing into the form of matter and energy of some mysterious panpsychist primordial Ur-consciousness that precedes spacetime. All of these ideas (or, in some cases it might be argued, *faux*-"ideas") have also been

D. Broderick, *Consciousness and Science Fiction*, Science and Fiction,
https://doi.org/10.1007/978-3-030-00599-3

brought on the intellectual playing field in science fictional or *fantastika* garb. And they have led here to a somewhat arbitrary gathering of stories and novels drawing on the topic of consciousness as a key driver of the fiction, not just as part of the inevitable background of any narrative.

Let us consider the narratives looked at in the body of this text, plus a few others, from several theorized standpoints (and by all means feel free to disagree with the following parsing):

One: consciousness is an evolved biological consequence of neural complexity—

1818/1831 Mary W. Shelley, *Frankenstein*
1939 Stanley G. Weinbaum, *The New Adam*
1948 Theodore Sturgeon, "Maturity"
1978 Whitley Strieber, *The Wolfen*
2000 Joan Slonczewski, *Brain Plague*
2002 Paul Levinson, *The Consciousness Plague*
2017 David Walton, *The Genius Plague*

One-prime: such consciousness could be created, evoked, shared or simulated—

1886 Robert Louis Stevenson, *Strange Case of Dr. Jekyll and Mr. Hyde*
1896 H.G. Wells, *The Island of Dr. Moreau*
1944 Olaf Stapledon, *Sirius*
1951 Wyman Guin, "Beyond Bedlam"
1960 James White, "Countercharm"
1963 Cordwainer Smith, "Think Blue, Count Two"
1976 John Crowley, *Beasts*
2014 Jo Walton, *My Real Children*
2017 Barbara Gowdy, *Little Sister*

Two: consciousness is an epiphenomenon or side effect of computations run in body and brain—

1992 Greg Egan, *Quarantine*
1994 Greg Egan, *Permutation City*
1997 Greg Egan, *Diaspora*
2010 Hannu Rajaniemi, *The Quantum Thief* (and sequels)

Two-prime: such consciousness could be created computationally in computers or robots—

1939–1986 Isaac Asimov, the Robots sequence
1947 Jack Williamson, "With Folded Hands"
1949 Ray Bradbury, "Marionettes, Inc."
1966 Robert Heinlein, *The Moon Is a Harsh Mistress*
1974 Barrington J. Bayley, *Soul of the Robot*
1980–1983 John Sladek, *Roderick* and *Roderick at Random*
1982 Tanith Lee, *The Silver Metal Lover*
1984 John Varley, "Press Enter ▮"
2009 Rachel Swirsky, "Eros, Philia, Agape"

Three: consciousness is an emergent form of matter and energy of a panpsychist primordial Ur-consciousness that precedes spacetime—

2017 Stephan A. Schwartz, *Awakening*

Three-prime: such consciousness could be found "embodied" in unexpected places, such as stars, or even everywhere, and as "life after death" either in a dimensional domain science has not yet accounted for or as the activating principle of a reincarnated person—

1937 Olaf Stapledon, *Star Maker*
1957 Fred Hoyle, *The Black Cloud*
1982–1997 Spider Robinson, *The Lifehouse Trilogy*
2000 Gregory Benford, *Eater*

Four: consciousness is an optional extra not found in philosophic zombies which are otherwise indistinguishable from conscious entities—

2016 Robert J. Sawyer, *Quantum Night*

Four-prime: consciousness is a user illusion and has no indisputable ontic reality—

1969 Philip K. Dick, *Ubik*
2006/2014 Peter Watts, *Blindsight* and *Echopraxia*

Five: consciousness is located in the brain, and can be transferred to another body or substrate—

1927 Edgar Rice Burroughs, *Master Mind of Mars*
1961–1969 Anne McCaffrey, *The Ship Who Sang*
1962 Brian W. Aldiss, “Shards”; 1962 “A Kind of Artistry”
1975 Lee Harding, *A World of Shadows*
2003 Justina Robson, *Natural History*
2005 Robert J. Sawyer, *Mindscan*
2014 John Scalzi, *Lock In*; 2018 *Head On*

Six: consciousness can be enhanced by brain modifications or mutations—

1939–1950 A.E. van Vogt, *Space Beagle* stories; 1942–1977 “Asylum” and extended as *Supermind*; 1948–1956 *The World of Null-A*, *The Pawns of Null-A*
1946 Lewis Padgett, *The Fairy Chessmen*
1950 Theodore Sturgeon, *The Dreaming Jewels*
1950 Katherine MacLean, “Incommunicado”
1954 Poul Anderson, *Brain Wave*
1968 Josephine Saxton, “The Consciousness Machine”
1983 Pamela Sargent, *The Golden Space*
1993 Nancy Kress, *Beggars in Spain* (and sequels)
2000 Jamil Nasir, *Distance Haze*

* * *

In a parallel volume of this Springer Series on Science and Fiction, Russell Blackford notes that “New understandings of the universe and ourselves, together with new technologies, have produced new philosophies and cultural movements” (*Science Fiction and the Moral Imagination* (2017, p. 178). He is reflecting on transhumanism and posthumanism, those recent innovative approaches to human involvement in the world of new techniques that press upon the role of consciousness. Some day, it seems likely, we shall confront true aliens, beings evolved (perhaps with guidance and selection from their ancestors) to experience the world in ways very different from ours. Even more probably, we shall share this planet with machine mentalities, “artificial general intelligences” that might be designed to complement our own evolved consciousness as partners or perhaps emerge from military laboratories as mechanisms of death and power.

These are potentials long ago dramatized by Philip K. Dick, whose crazed and astonishing futures from the 1950s and 1960s startlingly captured some of the changes we have already undergone in the twenty-first century and others we shall surely confront in coming decades. It was almost inevitable, looking in hindsight, that his zany, rich work would be pillaged for the mass media—sometimes to marvelous effect, as in *Blade Runner*, sometimes coarsely and witlessly, as with *Total Recall*.

Meanwhile, though, similar possibilities have been recycled intelligently for film and cable, in series such as *Person of Interest* (2011–2016) and the dream-sharing movie *Inception* (2010). One yearns for a truly sensitive and sensible enactment of, say, one of Greg Egan's brilliant stories or novels, or a Nancy Kress series such as the *Beggars* sequence, or Jo Walton's *My Real Children*, or even Theodore Sturgeon's *The Dreaming Jewels* or *More than Human*, difficult as those might be to deliver without trashing their nuance. Happily, these splendid works suffused by the problematics of consciousness remain available to be read—if not in paper, which seems to be falling out of favor as a medium, then as iPad-hosted ebooks or pocket-carried audio books.

It might even seem especially suitable that these marvels of printed literacy might find their most penetrating presence in forms beyond the imagination of the earliest pre-printing book copiers, painfully scratching each word with a quill on parchment. Arguably the ultimate fate for works considered here, prizing at the hinges of human and machine consciousness, will be experienced directly by upload into the brain—or the soul, or panpsyche, if either of those turns out to be the proper locus of the mind.

San Antonio, Texas, July 2018

References and Suggested Links

Science

Adam Becker *What Is Real? The Unfinished Quest for the Meaning of Quantum Physics* (Basic Books, 2018)
Russell Blackford *Science Fiction and the Moral Imagination* (Springer, 2017)
Russell Blackford, Van Ikin, and Sean McMullen: *Strange Constellations: A History of Australian Science Fiction* (Greenwood Press, 1999)
Susan Blackmore *The Meme Machine* (Oxford UP, 1999)
Damien Broderick *The Spike* (Reed [Aust.] 1997, revised Tor 2001)
Damien Broderick and Russell Blackford, eds *Intelligence Unbound: The Future of Uploading and Machine Minds* (Wiley Blackwell, 2014)
Damien Broderick and Ben Goertzel, eds *Evidence for Psi* (McFarland, 2015)
William Calvin *How Brains Think: Evolving Intelligence, Then and Now* (Weidenfeld & Nicolson, 1997)
William Calvin and Derek Bickerton *Lingua ex Machina: Reconciling Darwin and Chomsky with the human brain* (MIT Press, 2000)
Jeremy Campbell *The Improbable Machine: What the Upheavals in Artificial Intelligence Research Reveal about How the Mind Really Works* (Simon & Schuster, 1989)
David Chalmers *The Conscious Mind: In Search of a Fundamental Theory* (Oxford UP, 1996)
—— "Idealism and the Mind-Body Problem" (in W. Seager, ed., *The Routledge Companion to Panpsychism*, Oxford University Press, 2018; see http://consc.net/papers/idealism.pdf)
Patricia Churchland *Touching a Nerve: The Self as Brain* (Norton, 2013)
John Clute et al., *Encyclopedia of Science Fiction* (http://www.sf-encyclopedia.com/)

D. Broderick, *Consciousness and Science Fiction*, Science and Fiction,
https://doi.org/10.1007/978-3-030-00599-3

Francis Crick *The Astonishing Hypothesis: The Scientific Search for the Soul* (Simon & Schuster, 1994)
Alan Cromer *Uncommon Sense: The Heretical Nature of Science* (Oxford UP, 1993)
Rick Cytowic *The Man Who Tasted Shapes* (MIT Press, 2003)
Antonio Damasio, *The Feeling of What Happens: Body and Emotion in the Making of Consciousness* (Mariner Books, 2000)
—— *Descartes' Error: Emotion, Reason and the Human Brain* (Picador, 1995)
—— *Self Comes to Mind: Constructing the Conscious Brain* (Vintage, 2012)
Stanislas Dehaene *Consciousness and the Brain: Deciphering How the Brain Codes Our Thoughts* (Viking, 2014)
Daniel Dennett *Consciousness Explained* (Little Brown 1991)
—— *Darwin's Dangerous Idea: Evolution and the Meanings of Life* (Allen Lane, 1995)
—— *Kinds of Minds: Towards an Understanding of Consciousness* (Weidenfeld & Nicolson, 1996)
—— *Intuition Pumps* (Norton, 2013)
—— *From Bacteria to Bach and Back* (Norton, 2017)
K. Eric Drexler, *Engines of Creation: The Coming Era of Nanotechnology* (Doubleday, 1986)
Gerald Edelman *Bright Air, Brilliant Fire: On the Matter of the Mind* (Allen Lane, 1992)
Jason W. Ellis *Neuroscience, Science Fiction and Literature* https://dynamicsubspace.net/research/neuroscience-and-science-fiction-literature/
Todd E. Feinberg. M.D., *Altered Egos: How the Brain Creates the Self* (Oxford UP, 2001)
Edward Feser *The Last Superstition: A Refutation of the New Atheism* (St. Augustines Press, 2008)
Jerry Fodor *The Modularity of Mind* (MIT Press, 1983)
—— *Psychosemantics: The Problem of Meaning in the Philosophy of Mind* (Bradford Book, 1989)
Howard Gardner *The Mind's New Science: A History of the Cognitive Revolution* (Basic Books, 1986)
Michael Gazzaniga *Nature's Mind: The Biological Roots of thinking, Emotions, Sexuality, Language and Intelligence* (Penguin, 1994)
Susan Greenfield, *The Human Brain: A Guided Tour* (Weidenfeld & Nicolson, 1997)
J. Storrs Hall, PhD, *Beyond AI: Creating the Conscience of the Machine* (Prometheus Books, 2007)
Stuart Hameroff https://www.quantumconsciousness.org/content/overview-sh
Nicholas Humphrey *A History of the Mind* (Chatto & Windus, 1992a)
John Horgan *The Undiscovered Mind: How the Human Brain Defies Replication, Medication, and Explanation* (The Free Press, 1999)
Jeff Hawkins and Sandra Blakeslee *On Intelligence* (Times Books/Henry Holt 2004)
Douglas Hofstadter *Gödel, Escher, Bach: An Eternal Golden Braid* (Penguin, 1980)

—— and Daniel Dennett, eds, *The Mind's I: Fantasies and Reflections on Self and Soul* (Penguin, 1981)
Nicholas Humphrey *A History of the Mind* (Chatto & Windus, 1992)
Ray Jackendoff *Semantics and Cognition (MIT Press, 1983)*
—— *Consciousness and the Computational Mind (MIT Press, 1987)*
Philip N. Johnson-Laird *Mental Models: Towards a Cognitive Science of Language, Inference, and Consciousness* (Harvard UP, 1983)
Jerome Kagan *Galen's Prophecy: Temperament in Human Nature* (Basic Books, 1994)
Arthur Koestler *The Act of Creation* (Hutchinson, 1964)
Ray Kurzweil *How to Create a Mind: The Secret of Human Thought Revealed* (2012).
Chris Langton, cited in Waldrop (1992).
John McCrone *The Myth of Irrationality: The Science of the Mind from Plato to Star Trek* (McMillan, 1993)
Marvin Minsky *The Society of Mind* (Picador, 1988)
Thomas Nagel "What Is It Like to Be a Bat?" (reprinted in Hofstadter and Dennett, eds, *The Mind's I*, pp. 391–403)
Alexei Panshin and Cory Panshin *The World Beyond the Hill: Science Fiction and the Quest for Transcendence* (Tarcher, 1989)
Roger Penrose *The Emperor's New Mind: Concerning Computers, Minds, and the Laws of Physics* (Oxford UP, 1989)
—— *Shadows of the Mind: A Search for the Missing Science of Consciousness* (Oxford UP, 1994)
Steven Pinker *The Language Instinct: The New Science of Language and Mind* (Allen Lane, 1994)
—— *How the Mind Works* (Allen Lane, 1997)
—— *The Blank Slate: The Modern Denial of Human Nature* (Allen Lane, 2002)
Dean Radin *Real Magic: Ancient Wisdom, Modern Science, and a Guide to the Secret Power of the Universe* (2018)
V.S. Ramachandran and Sandra Blakeslee, *Phantoms in the Brain: Human Nature and the Architecture of the Mind* (Fourth Estate, 1998)
Ken Richardson *The Making of Intelligence* (Weidenfeld & Nicolson, 1999)
Sue Savage-Rumbaugh and Roger Lewin, *Kanzi: The Ape at the Brink of the Human Mind* (Doubleday, 1994)
Maurice Schouten and Huib Looren de Jong, eds. *The Matter of the Mind: Philosophical Essays on Psychology, Neuroscience, and Reduction* (Wiley-Blackwell, 2012)
Stephan A. Schwartz Six Protocols, Neuroscience, and Near Death: An Emerging Paradigm Incorporating Nonlocal Consciousness *Explore*: Volume 11, Issue 4, July–August, 2015, pp. 252–60 https://www.explorejournal.com/article/S1550-8307(15)00076-2/fulltext
John R. Searle *Consciousness and Language* (Cambridge UP, 2002)
Max Tegmark *Life 3.0: Being Human in the Age of Artificial Intelligence* (MIT Press, 2017)

Corbett H. Thigpen and Hervey M. Cleckley *The Three Faces of Eve* (McGraw-Hill, 1957)

Giulio Tononi *Phi: A Voyage From the Brain To the Soul* (Pantheon, 2012)

—— and Christof Koch, 2015, Consciousness: Here, There and Everywhere? *Philosophical Transactions of the Royal Society B: Biological Sciences,* 370(1668): 20140167. doi:https://doi.org/10.1098/rstb.2014.0167

Zoltan Torey *The Crucible of Consciousness: A Personal Exploration of the Conscious Mind* (Oxford UP, 1999)

Mitchell M. Waldrop *Complexity: The Emerging Science at the Edge of Order and Chaos* (Simon & Schuster, 1992)

Eugene Webb, *Philosophers of Consciousness: Polyani, Lonergan, Voegelin, Ricoeur, Girard, Kierkegaard* (University of Washington Press, 1988)

Gary Westfahl *The Spacesuit Film: A History* (McFarland, 2012)

—— *Islands in the Sky: The Space Station Theme in Science Fiction Literature* (Borgo Press, 1996, 2009)

Caroline Williams *My Plastic Brain: One Woman's Yearlong Journey to Discover if Science Can Improve Her Mind* (Prometheus, 2017)

Lisa Zyga https://phys.org/news/2009-06-quantum-mysticism-forgotten.html (2009)

Fiction

1962 Brian Aldiss, "Shards" and "A Kind of Artistry"
1954 Poul Anderson, *Brain Wave*
1939–1950 Isaac Asimov, the Robots sequence
1949 Ray Bradbury "Marionettes, Inc."
1927 Edgar Rice Burroughs, *Master Mind of Mars*
2015 Brenda Cooper, *Edge of Dark*
2016 —— *Spear of Light*
1976 John Crowley, *Beasts*
1969 Philip K. Dick. *Ubik*
1992 Greg Egan, *Quarantine*
2017 Barbara Gowdy, *Little Sister*
1951 Wyman Guin, "Beyond Bedlam"
1975 Lee Harding, *A World of Shadows*
1966 Robert A. Heinlein, *The Moon is a Harsh Mistress*
1993 Nancy Kress, *Beggars in Spain*
2002 Paul Levinson, *The Consciousness Plague*
1961–69 Anne McCaffrey. *The Ship Who Sang*
1950 Katherine MacLean, "Incommunicado"
1998 Ken MacLeod *The Cassini Division*
2000 Jamil Nasir, *Distance Haze*
1946 Lewis Padgett, *The Fairy Chessmen* aka *Chessboard Planet*

1991 Charles Platt, *The Silicon Man*
1982–97 Spider Robinson, *Lifehouse Trilogy* (2007), comprising: *Mindkiller* (1982), *Time Pressure* (1987), *Lifehouse* (1997)
2005 Justina Robson, *Natural History*
1982 Pamela Sargent, *The Golden Space*
2005 Robert J. Sawyer, *Mindscan*
2016 —— *Quantum Night*
1968 Josephine Saxton, "The Consciousness Machine"
1818 Mary Shelley, *Frankenstein, or, The Modern Prometheus* (Project Gutenberg https://www.gutenberg.org/files/84/84-h/84-h.htm)
1953 Curt Siodmak *Donovan's Brain*
2000 Joan Slonczewski, *Brain Plague*
1963 Cordwainer Smith, "Think Blue, Count Two"
1944 Olaf Stapledon, *Sirius*
1886 Robert Louis Stevenson, *Strange Case of Dr. Jekyll and Mr. Hyde* (Project Gutenberg EBook http://www.gutenberg.org/ebooks/43)
1978 Whitley Strieber, *The Wolfen*
1948 Theodore Sturgeon "Maturity"
1950 Theodore Sturgeon, *The Dreaming Jewels*
2009 Rachel Swirsky, "Eros, Philia, Agape"
1991 George Turner, *Brain Child*
1939–1950 A.E. van Vogt, *Space Beagle* stories; 1942–77 "Asylum" and extended as *Supermind*; 1948–56 *The World of Null-A, The Pawns of Null-A*.
1984 John Varley, "Press Enter ▮"
2017 David Walton, *The Genius Plague*
2014 Jo Walton, *My Real Children*
2006 Peter Watts, *Blindsight*
2014 —— *Echopraxia*
1939 Stanley G. Weinbaum, *The New Adam*
1896 H.G. Wells, *The Island of Dr. Moreau*
1960, 1969 James White, "Countercharm"
1947 Jack Williamson, "With Folded Hands"

Index

D. Broderick, *Consciousness and Science Fiction*, Science and Fiction,
https://doi.org/10.1007/978-3-030-00599-3